BEHIND THE SCIENCE

The Invisible Work of Data Management in Big Science

Katherine Harrison

First published in Great Britain in 2025 by

Bristol University Press
University of Bristol
1–9 Old Park Hill
Bristol
BS2 8BB
UK
t: +44 (0)117 374 6645
e: bup-info@bristol.ac.uk

Details of international sales and distribution partners are available at bristoluniversitypress.co.uk

British Library Cataloguing in Publication Data
A catalogue record for this book is available from the British Library

ISBN 978-1-5292-3010-9 paperback
ISBN 978-1-5292-3011-6 ePub
ISBN 978-1-5292-3012-3 OA PDF

Cover design: blu inc
Front cover image: Alamy/Iakov Filimonov
Bristol University Press uses environmentally responsible print partners.
Printed and bound in Great Britain by CPI Group (UK) Ltd, Croydon, CR0 4YY

FSC
www.fsc.org
MIX
Paper | Supporting responsible forestry
FSC® C013604

Contents

List of Figures

About the Author

Katherine Harrison is Associate Professor in Gender Studies at Linköping University, Sweden. Her research sits at the intersection of Science and Technology Studies, media studies, and feminist theory, bringing critical perspectives on knowledge production to studies of different digital technologies. She has received funding from the Swedish Civil Contingencies Agency, the Danish Council for Independent Research, Marcus and Amalia Wallenberg Foundation, Riksbankens Jubileumsfond, Marianne and Marcus Wallenberg Foundation and the Swedish Research Council for Sustainable Development (FORMAS). She is currently co-PI for two WASP-HS (Wallenberg AI, Autonomous Systems and Software Program – Humanity and Society) projects (*The ethics and social consequences of AI and caring robots* and *Operationalising ethics for AI*).

Acknowledgements

This research was supported financially by several different funding bodies at different stages in its life. My thanks go to the Marcus and Amalia Wallenberg Foundation for providing me with funding for the original postdoctoral project (grant number MAW – 2014.0033). Towards the end of the fieldwork period, I was invited to join a Lund University-supported interdisciplinary project at the Pufendorf Institute titled 'DATA: Enabling us to better store, observe and understand what we measure' led by Melvyn B. Davies, Monica Lassi and Kalle Aström. The conversations in this DATA group played a significant role in developing the theoretical framing for this book. Later, in the writing stage, I benefited from a writing retreat supported by the Department of Thematic Studies at Linköping University and most recently from a sabbatical grant from Riksbankens Jubileumsfond (grant number SAB22–0063).

In this project I was in the fantastically lucky position of being able to interview people who are experts in their fields, passionate about what they do and with whom it was a pleasure to talk. The conversations I had with Jon, Tobias, Thomas, Sune, Afonso, Petra and Mark were educational and thought-provoking. We also laughed sometimes as well, as I bumbled my way through trying to make sense of the extremely complex technical work they do. Thank you all for your time and patience! In this book I have tried to do justice to your perspectives while respectfully integrating my critical scholarly voice – I hope you enjoy reading it.

I'd like to also extend my thanks to the people I met at the Diamond Light Source (Silvia da Graca Ramos and Claire Murray), at STFC/ISIS (Tom Griffin, Hannah Griffin, Matt Clarke, Stephen King and Kevin Phipps) and at MAX IV (Darren Spruce and Vincent Hardion). My visits to these facilities were an important part of my learning experience – thank you very much for your time and hospitality.

This book took shape in Crete during 2023 when I spent six months on sabbatical at the Institute of Computer Science, at the Foundation for Research and Technology (FORTH), in Heraklion, Crete, Greece. There are many people at FORTH to thank for making my time there so rewarding, but particular thanks go to my main host, Professor Angelos Bilas. Eleni

Pavlopoulou and Fotis Nikolaidis were kind enough to read and comment on early drafts of some of the main chapters – our conversations were one of my favourite parts of being at FORTH! Meanwhile, Maria Prevelianaki and Alison Manganas helped me to navigate the everyday challenges with enormous patience and good humour. A warm thank you also goes to my in-laws, Giorgos and Nafsika Angelakis, who provided indispensable support at home and with the kids during the sabbatical so that I could focus on writing.

Paul Stevens at Bristol University Press has been a calm quiet presence throughout the whole process – thank you Paul. I am also grateful to the reviewers of the original book proposal for their encouraging feedback. I am fortunate to be part of a lively Science and Technology Studies (STS) community in Sweden, which has provided feedback in various forms including conferences and seminars. I'd like to say a special thank you to Corinna Kruse, Francis Lee, Kerstin Sandell, Julia Velkova and Charles Berrett for reading and commenting on chapter drafts during the later stages.

My colleagues at Tema Genus have been both patient and kind when I have disappeared into my 'book cave' to write – your collegial support is a real blessing. However, particular mention has to go to Ericka Johnson for support/amusing email exchanges/encouraging me to write the book in the first place, and Alma Persson for making it possible for me to go on sabbatical.

Somewhere, 'behind the book' are the folks whose support is a little harder to see, but is just as essential. My parents, Jenny and David Harrison, were always there. And, last but not least, I would have become a total data nerd during the writing of this book if it wasn't for Vangelis, Dominic and Artemis regularly demanding that I go to the beach, eat ice cream, discuss dinosaurs and other important life stuff. Thank you!

1

Data and Knowledge

In a field outside Lund in southern Sweden, the world's most powerful neutron source is finally ready for action. It may sound like the stuff of science fiction, but in fact it is a very real tool that may well help us to understand another of the universe's mysteries or solve a major societal challenge. At least, that is the promise for the future (Dimitrievski, 2019). After all, how else could spending more than 2 billion euros and 10 years under construction be justified?[1]

This neutron source is located at the heart of the European Spallation Source (ESS), a Big Science facility that – when fully operational – will produce some of the largest quantities of data in today's data-dominated world. Scientists from all over the world will travel to this experimental facility to test samples, hoping for cutting-edge discoveries like those that have happened at similar sites such as CERN.[2] The facility is scheduled to be fully operational by 2027[3] when an estimated 2,000–3,000 visiting scientists will come every year to conduct experiments. Reliable capture and management of such experimental data is essential to ensure that scientific results have a solid foundation. However, the visibility of those working with data management and the institutional support for such activities at such facilities has historically been low. This book starts by asking two simple questions: if data is so fundamental to scientific results achieved at these facilities, why has data management not been more visible both in scholarly accounts of these facilities and in the facilities themselves?

1 The initial cost estimate was 1.84 billion euros (https://www.lunduniversity.lu.se/research-innovation/max-iv-and-ess). In late 2021, an additional cost of 550 million euros was approved to reflect delays incurred due to the COVID-19 pandemic (https://europeanspallationsource.se/article/2021/12/10/ess-revises-project-plan-and-budget).

2 CERN is a Big Science facility located in Geneva, perhaps most famous in recent years for the discovery of the Higgs boson.

3 European Spallation Source 2022 activity report (https://europeanspallationsource.se/sites/default/files/files/document/2023-06/ESS0064_A%CC%8Arsbok_2022_screen.pdf).

And what does this matter to how knowledge is produced at Big Science facilities like the ESS?

To answer these questions, this book tells the story of a unique research journey following the people responsible for designing and implementing data management at the ESS. It explores how decisions about data management occur as the result of complex intersections of people, technologies and organisational politics. As such, it aims to highlight the context in which data is produced by following the trail of decisions made about data management during the construction of a Big Science facility. It draws on several years of contact with technical experts responsible for designing these systems who were patient enough to explain their working practices and personal experiences in these highly specialised facilities to a curious social scientist.

This is a story with a lot of movement in it. Movement in terms of the facility itself, whose foundations were literally being dug when I first met my participants, and where I watched the plans for data management evolve from sketches to realities during the course of my fieldwork. In contrast to the visits I made to other Big Science facilities which often started with a guided tour of the experimental areas and server rooms, conducting research at the ESS mostly involved interviews conducted in anonymous offices in temporary premises. This meant that there was also significant movement in terms of my own expectations of the field, as I started fieldwork fresh from reading the 'biographies' of other Big Science facilities awash with the rhetorical promise of world-changing discoveries, only to find myself interviewing participants contemplating a site that was a hole in the ground and with no server room. Looking at the bigger picture, as a pan-European project, this is also a facility built on international collaboration involving movement of equipment and knowledge across borders. Finally, following the data as it moved from detector to scientist became an important aspect of my study. All these different kinds of movement and change will be important in what follows.

Movement was also key to making the work done by these data management experts more visible. I am interested in metaphors of visibility and transparency in part because this is a scientific endeavour which is opaque to many people, but also because I sense that data management constitutes a kind of paradox in the 'normal' functioning of science. Reliable, reproducible results are premised on a certain amount of transparency; by describing the steps taken in an experiment it should be possible for someone else to reproduce the work and get the same results. However, asking how decisions about which software and hardware are used to perform data capture and management is an area normally placed firmly into a closed black box outside the scope of The Experiment itself. This makes this also a story about changing understandings of what Big Science is, and its role in society.

Part of this is the changing role of the technical experts that I met, whose skills are increasingly in demand to manage larger and more complex data sets. What is their role in the new order: scientist or technician? But before we set off, it could be helpful to know a little about the 'backstory' to the ESS and Big Science itself.

The construction team broke ground on this new Big Science facility in 2014 after more than 20 years of planning (Hallonsten, 2012; Kaiserfeld and O'Dell, 2013). Part of the reason for the long period of planning was the collaborative nature of the ESS. It is a pan-European project involving 13 partner nations[4] and based in Sweden and Denmark. Financially, Sweden and Denmark are the primary contributors to the project, resulting in the decision to locate the main facility in Lund, Sweden, and the Data Management and Software Centre (DMSC) in Copenhagen, Denmark. But what will scientists actually be doing at this facility?

> Neutron scattering allows scientists to count scattered neutrons, measure their energies and the angles at which they scatter, and map their final positions. This information can reveal the molecular and magnetic structure and behaviour of materials, such as high-temperature superconductors, polymers, metals, and biological samples. (Wikipedia, 2023)

Counting of the scattered neutrons takes place when – after contact with an experimental sample – the neutrons' path is registered by highly sensitive detectors surrounding the sample. Based on information such as the speed and angle of the neutrons, important insights into the experimental sample can be obtained. This information about the neutrons is the 'raw' data from the experiments which is collected and processed by the DMSC of the ESS and supplied to the visiting scientists.

Figure 1.1 is an illustration provided on the ESS website which aims to explain how an experiment will take place at this facility. On the far left-hand side of the illustration, there is the source that produces neutrons which travel at high speed towards an experimental sample. As these neutrons collide with the sample they are changed (for example, by losing energy or changing direction of travel). After colliding with the sample, the neutrons are registered by the detectors that surround the sample. From this data (numbers, energy, angle, and so on of the neutron) much can be learnt about the properties of sample.

---

[4] Map of European Spallation Source In-Kind partners (https://europeanspallationsource.se/sites/default/files/images/media/2023-12/ESS%20IK%20partners%20map%20and%20list_new.png).

Figure 1.1: The European Spallation Source infographic

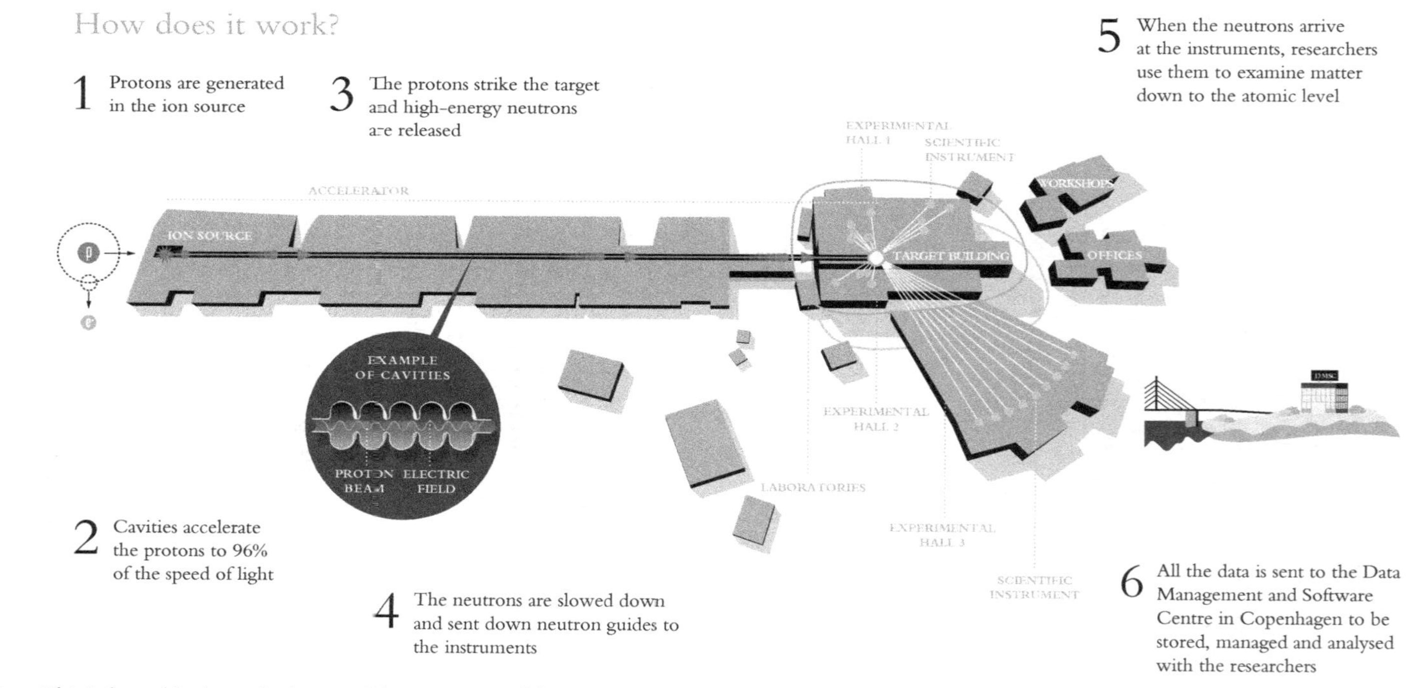

Note: This infographic shows the layout of the experimental facility in Lund, with a focus on the beamline, which is the horizontal line through the middle of the building marked with arrows.

Source: Courtesy of ESS, and available on the ESS website

However, between the detector and the visiting scientists is a complex array of procedures and technologies that move, reformat, clean, categorise and archive the data. This part of the work of Big Science didn't appear on any diagrams about how the ESS functions, nor is it included in any of the scholarly accounts of similar facilities.[5] It is, however, exactly this part of the ESS with which this book is concerned. Invisible to the external world, hidden by lack of awareness, as well as the promise of raw data being provided in real time to the experimenters, it is easy to overlook the essential work being done at this stage to provide data. Or, at least, to fail to appreciate the role of such work (performed both by people and technologies) in the experimental procedure. Indeed, it could even be argued that this lack of visibility serves an important purpose. What if the very existence of this work potentially troubles the 'rawness' of the data and needs to remain invisible so that the data may be considered valid, objective, and truly 'raw' when it arrives on the screens of the scientists?

In brief then, this book sets out to examine the changing nature of Big Science, critically investigate the foundational premise of data as Truth upon which experiments conducted at these facilities rely, and shed light on the undervalued yet essential work done by those designing and developing data management systems. It does this through a close study of a Big Science facility called the ESS during its early construction period. It will offer important empirical and theoretical contributions to the fields of Science and Technology Studies (STS) and Critical Data Studies (CDS), which comprise its primary audiences. Empirically, it provides a rare glimpse into the development of a Big Science facility, and the only one to date that sheds light on the development of data management systems. Theoretically, it adds to the ongoing scholarly work that explores how particular conceptualisations of 'data' are implicit in the production of 'scientific knowledge' by providing an example of a less-studied context in which tensions between the role of technical expert and scientist are particularly in focus.

In what follows, I will set the scene by outlining what is meant by the term 'Big Science' and how the ESS relates to this category. Then I will introduce the notion that data management work done at such facilities can be described as 'invisible work', before making an argument for why it should receive more attention. This is followed by introduction of the key

[5] The version of the illustration shown here was provided to the author by the facility during the late stages of production of this book. Unlike earlier versions of the illustration that appeared on the ESS website, it shows the Data Management and Software Centre and provides the reader with basic information about what happens there.

theoretical concepts that weave throughout the book and an overview of the chapters that follow.

From Big Science to New Big Science

Historically, Big Science was the result of research efforts during the Cold War typically realised in the form of research infrastructures like CERN, large-scale facilities where economies of scale and scope required extremely big sites. Today, this old version of Big Science remains a prestigious activity, albeit increasingly marginalised in the overall mix of scientific activities and in high-energy physics in particular. Scholars have instead sought to distinguish a new model termed 'New Big Science' (NBS). This, in contrast to the old Big Science which characterised the cold-war efforts, is less costly, dependent upon international collaboration, multidisciplinary, and enmeshed in social and economic processes with promises of contributions to scientific development, innovation and eventually economic growth (Hoddesdon et al, 2019; Rekers and Sandell, 2016). New Big Science facilities, such as the ESS[6] or MAX IV,[7] counter their cost to build and maintain by providing significant societal and intellectual impact in the form of jobs, scientific discoveries and new academic models of knowledge production. However, experimentation in these facilities is an increasingly advanced and complex process, requiring significant experience in terms of knowing what questions to ask, how to handle the instruments and manage the data stream.

In Big Science facilities, visiting scientists from all over the world use a range of advanced and unique instruments for experimentation collected around the central source, producing an exponentially increasing amount of data to be processed. The sophistication of the instruments, in combination with the large amounts of data, is fundamentally changing the technical and computational expertise needed to be able to produce experimental knowledge. A broader user group will also result in less experienced users visiting the facility, thus driving a demand for more onsite assistance in collecting, processing and analysing the data. Increasingly sophisticated digital technologies are essential for managing the new challenges and complexities associated with multidisciplinary data management. This changing nature of scientific enquiry is vital to understand as it has significant societal and intellectual impacts.

[6] https://europeanspallationsource.se/

[7] https://www.maxiv.lu.se/ The Swedish national Big Science facility, also located just outside Lund.

This book is concerned with data management at Big Science *facilities*, the epitome of what Derek de Solla Price described at the start of *Little Science, Big Science* as 'the large-scale character of modern science, new and shining and all-powerful' (1963, 2; see also Weinberg, 1961). While de Solla Price goes on to trouble any neat distinction between these modes of science in the course of his book, he does concede that one of the things that marks out 'Big Science' (and which remains true even today) is money (de Solla Price, 1963, 92). Big Science facilities such as the ESS are – in essence – some of the world's most expensive laboratories generating and processing the largest quantities of data in real time. The challenges and innovations connected to such facilities are technical, organisational and scientific. It is this distinctive nature of the facilities that therefore makes them such a fascinating case study in terms of scientific practices around data.

A glance at the scholarship shows how far the notion of 'Big Science' has come in the intervening 60 odd years since Weinberg and de Solla Price (see Cramer and Hallonsten, 2020 for a useful guide to how 'big science' relates to changing notions of research infrastructures). The term is often now associated with any kind of scientific endeavour involving large, geographically distributed research teams producing or curating large data sets, and not necessarily tied to a single facility or institution (Edwards, 2010; Vermeulen, 2013; Vermeulen, 2016; Scroggins and Pasquetto, 2020).[8] The Human Genome Project (HGP), for example, is often described as 'big science' by scholars because it involved advances both in understandings of biology and in technical resources. Writing in 'Big science and big data in nephrology', for example, Saez-Rodriguez et al describe 'big science' as 'the joint effort of large consortia to generate big data to help reach a common goal, and discuss how this can have a profound impact in nephrology' (Saez-Rodriguez et al, 2019, 95). Similarly, Niki Vermeulen makes a case for '(i)nvestigations into the oceans and their living creatures is big science avant la lettre' (2013, 3) by reconstructing a history of marine biology research that highlights the features that would categorise it as 'big science'. Climate change too has been argued for as a kind of 'big science' for some time now (Stanhill, 1999; Edwards, 2010). What then are the defining characteristics of these newer emergences of 'big science'?

Discussing the relation between big science and individual-investigator-oriented science in relation to the HGP, Hood and Rowen distinguish big science as providing 'resources that are foundational for all researchers' (2013, 2)

---

[8] These shifts in understanding have resulted in some scholars referring to 'big science' rather than 'Big Science' in efforts to distinguish between older and newer understandings of the term.

and refer instead to a 'big science approach' (Hood and Rowen 2013, 1). Complexity is also a term that is often associated with 'big science' where the problem under consideration is of a sufficiently large scope to require coordination across many teams (Hood and Rowen 2013). Calvert (2013), meanwhile, argues that 'big science' is the language of the twentieth century and is no longer a good fit with major scientific projects. Instead, she advocates for thinking in terms of 'grand challenges' where humanities and social sciences should also be included. Her twist on the term highlights the implicit 'hard science' bias of the term 'big science' as well as the need for grounding such major endeavours in societal realities. Calvert's description of big science as 'a policy category developed in a post-war funding environment and centred on physics (Smith, 1992)' (2013, 467) seeks to distance big science from the 'real work' of science, framing it as a strategic, political approach. Ultimately though, Calvert argues that whichever term is used it essentially indicates research priorities as evidenced through major investment.

This book, however, is focused on the workings of a Big Science *facility*.[9] The experimental sample is in a particular place and must be in proximity to the source for the experiment to take place. The detectors are at a prerequisite distance from the sample, ready to collect data about the sample. The data travels from detectors in Lund to a server room in Copenhagen down a dedicated pipe. These are physical coordinates that work to make the ESS the eye of the storm or the centre of the story. But the journey taken by the data is surrounded by other kinds of journeys, for example the transfer of expertise from 'In-Kind Contribution' (IKC) partners to DMSC core team. Money, expertise, people and equipment flow into the ESS. Data flows out. But at the start of the data journey, for less time than the blink of an eye, the data is at the ESS, making its way from detector to screen, undergoing cleaning and aligning in real time. And the hardware that supports this is at the ESS. The people who planned and built this system are at the ESS, and they are at the heart of this book.

The ESS is a single facility, with a huge price tag and surrounded by visionary rhetoric. It will provide experimental facilities to both the 'traditional' Big Science communities, such as High Energy Physics (HEP) but equally to other communities from Life Sciences, Materials Science and even the Humanities. This book offers a glimpse into the internal workings of the facility, where people and technologies are entangled and where decisions about data management are shaped by a distinctive organisational context. The focus then here is not on a time-limited project situated within a particular discipline or topic and organised around a research question but

[9] And here I deliberately capitalise Big Science as a way to indicate that I am referring to the older definition of the term grounded in a single facility.

on an ongoing, long-term physical facility staffed by technical experts focused on high-quality data production for a wide range of scientific communities. It is not a study of the specific practices or cultures seen among scientists using these facilities (for example, Cetina, 1999) but on the technical and organisational challenges involved in selecting, connecting, and aligning the software and hardware that must deliver the brand new 'valid' data (Kruse, 2006) that is an essential basis for scientific breakthroughs.

The distinction between little and big science that de Solla Price unpicked in his 1963 volume is often cited by data scholars describing differences in data sets. For example, Christine Borgman's 2015 volume *Big Data, Little Data, No Data* uses this an opening gambit, noting that: '(j)ust as big science was to reveal the secrets of the universe, big data is expected to reveal the buried treasures in the bit stream of life' (3). What both Big Science and big data have in common is a tendency towards visionary rhetoric that promises definitive answers to society's big questions. However, critical studies of big data by communication and digital media scholars such as Borgman (2015), Gitelman (2013), Hine (2006) and Kitchin (2014) have drawn attention to the ways in which decisions about data management and virtual infrastructure constitute a form of knowledge production. These often draw upon STS concepts such as 'situatedness' to stress that 'how data are conceived, measured and employed actively frames their nature' (Kitchin and Lauriault, 2014; see also Ribes, 2019, and Gillespie, Boczkowski and Foot, 2014). Critical studies such as these have already provided insights into a range of different empirical field sites of data production, but as yet they have paid little attention to the context of Big Science facilities.

The changing face of data management

Big Science facilities like the ESS have received attention from numerous STS scholars who have explored both the creation of specific facilities (for example, Doing, 2009; Hoddesdon et al, 2019) and the HEP scientific communities that – historically at least – have been the main users of such facilities (for example, Pickering, 1986; Traweek, 2009). This scholarship describes in detail the distinctive practices and scientific cultures that develop around these facilities but remains relatively silent on the supporting infrastructure such as computing or data management. One of the most famous contributions of this latter category, *Image and Logic, a Material Culture of Microphysics* by Peter Galison (1997), touches briefly on computing but does not discuss it at length. This is representative of a general absence of scholarship concerning the role of what has historically been referred to as 'scientific computing' as part of the STS histories of Big Science. What little there is tends to place data management in a supporting role to the 'main job' of Big Science, that is, conducting experiments (Seidel, 2008).

In 'From Factory to Farm: Dissemination of Computing' Robert W. Seidel narrates the integration of super-computers into HEP from the 1960s to the 1980s (2008). In it he cites a 1985 report about future computing needs for particle physics that is particularly telling. The report makes it clear that scientific computing within HEP had been neglected to date and was in need of more attention: 'it would be a bad mistake to follow the usual path of machine first, detector second and computing last. It could easily be that the computing problem for this class of machine and detector may present the biggest intellectual challenge of all' (from the 'Computing for Particle Physics: Report of the HEPAP Subpanel on Computer Needs for the Next Decade', TR DOE/ER-0234 (Washington, DC: DOE Office of Energy Research, 1985),14, as it appears in Seidel, 2008, 504).

Seidel describes how the developments in HEP (for example, more sensitive detectors that allowed more data to be captured or captured data from a bigger area) led to the production of an increasing amount of data, thereby creating a need for more efficient or different ways of managing data. HEP developments thus stimulated developments in data management tools, which then required scientists to develop further their own computing skills. This was not a new process; developments in physics had often produced a need for new skills. Seidel cites the histogram, for example, as being one of these moments: '"High-statistics physics" reified data from many events detected by bubble and spark chambers and their more sophisticated descendants into histograms – statistical frequency analyses of their properties. Because this was less straightforward than interpreting a photograph, *physicists had to be trained to decipher them*' (2008, 485, my emphasis).

However, Seidel's narrative also makes clear how different tensions and local constraints shaped the speed at which individual institutions acquired and integrated supercomputers into their experimental set-ups. He highlights not only the demands from scientists but also the prohibitive costs associated with such machines and the personal opinions of key institutional decision-makers with regards to use of computers for data management. Seidel also notes the move from software designed in-house to the use of external firms, leading to a long-term relationship with and dependence upon commercial software houses. This outsourcing could be understood as a reflection of the Big Science facilities' choice at that time to focus their skills, energy and budget on the experiments rather than computing. Computing (in the form of both hardware and software) thus appears to have – historically at least – been considered sufficiently separate from the knowledge production process that it could be outsourced in this way.

Fast forward to the 2000s and, from the very start, plans for the creation of the ESS included a specially dedicated DMSC to be built in Copenhagen. This centre described its mandate as follows:

> The Computing Centre at DMSC is currently in both an operational and a design phase, and its mission is twofold:
>
> - To provide scientific computing services and computing capacity (High Performance Computing, HPC) to the ESS divisions in the planning, construction and commissioning phases of the ESS facility in Lund.
> - To develop, design and build expertise and systems in areas relevant required to support the scientific use case once ESS becomes operational. (ESS Computer Centre, 2024)

As the quotation from the ESS website suggests, this data management centre is framed (to a large extent) as supporting act to the 'main show' of the science being conducted on the instruments. This looks familiar from Seidel's account of earlier computing developments in Big Science. The creation of a dedicated data management centre also clearly shows that handling the large amounts of data produced by experiments remains an ongoing challenge for Big Science facilities.

However, the case of the DMSC also highlights some interesting changes in the understanding of data. First, data management now constitutes a visible and valued part of the facility. This is evident in several ways, including how the existence of and need for the DMSC is visible and acknowledged through its presence on the ESS website, where it is listed in the section of the ESS website called 'Science and Instruments'. Its positioning there frames it as an integral part of the facility and the process of doing science. The DMSC staff (rising from 11 people as of May 2015 to nearly 30 in Fall 2022) are organised into five teams (comprising; Data Systems and Technologies, Data Management, Instrument Data, Data Analysis and Modelling, and Scientific Coordination and User Office). This pool of experts with different specialisations suggests a significant level of investment in management of data at the ESS, and a greater recognition of the role played by data acquisition and analysis in the knowledge production process compared to the approach that Seidel outlines from the 1960s to 1980s. It is also worth noting that the choice of location for the DMSC constitutes a public acknowledgement of the involvement of one of the major funding partners, Denmark. The centre itself can thus be understood as sufficiently important within the whole project to function as an appropriate reflection of a major financial contribution to the overall project (although still as a part of the ESS that can be – more or less – easily detached from the rest of the facility).

The experts you will meet in this book are based at the DMSC of the ESS. The DMSC has a clear brief to provide information and communications technology (ICT) support to the ESS as well as handling operation of the computing cluster and providing 'infrastructure services'. These responsibilities require internal and external cooperation with researchers, software developers and facility management; organising how to collect, move, analyse or store data is a collaborative enterprise that shapes organisational infrastructure in the form of meetings, budgetary considerations, working agreements and logistics. Separation of the two sites has required, for example, the implementation of 'a redundant dedicated fibre connection' (ESS website, Data Management), increased use of video conferencing tools and a longstanding commitment by DMSC staff to divide their working weeks between the two sites. All of these make concrete demands on organisational infrastructure in terms of money, time and space. These tangible resources reflect a growing recognition that technical support around data management is going to be necessary, that it is no longer a case of researchers being able to handle their own data.

In/visible infrastructures

In light of the previous discussion, it is more important than ever to ask: why is data management overlooked in studies of Big Science facilities, and what does it matter specifically to knowledge production in Big Science? To answer these questions, this book follows the construction of the ESS, and more specifically the process of designing and developing data management systems as experienced by the Group Leaders of the DMSC. It looks at how organisational politics, personal experience and material limitations of the technologies intersect in an organisation that is under construction.

In order to explore the in/visibility of the work carried out at the DMSC, I have drawn on a number of well-developed tools from the STS literature, namely black boxes and invisible work. In *Science in Action: How to Follow Scientists and Engineers Through Society*, Bruno Latour suggests that technologies or facts can become black boxes, self-contained, opaque entities apparently without history or controversy (1987). This can make it difficult to examine critically a technology. It can be hard to imagine how user representations, technical limitations or organisational priorities shaped a technology when one is standing in front of the finished product. Part of the uniqueness of this book therefore lies in the opportunity to see 'behind the scenes' or more precisely 'behind the science' by virtue of being there while the ESS is under construction. Not only do we gain the chance to 'be there before the box is closed' (Latour, 1987, 21), but also this book constitutes a contribution to accounts of construction of Big Science facilities.

Approaching the data management system of the ESS as a series of 'black boxes' which are not yet closed makes it possible to ask critical questions about the way in which these technologies are being developed, and to explore the consequences of that. I argue that during construction it is easier to see the 'invisible work' of developing data management systems. Following the development of data management systems at the ESS as they progress from back-of-the-envelope sketches to tangible hardware solutions allows us to see how particular technical solutions emerge from specific contexts of use and experience. Not only does it make the logic for choosing one technical solution more visible but also the work done by the data management group and their international collaborators to ensure reliable collection and delivery of experimental data. As such, I also take inspiration from Star and Ruhleder's note that invisible work is easier to see when things are broken (1994) and not working perfectly. Their work has been used extensively in studies of infrastructure as a way to lift up processes and practices which often seem to take place in the background, yet which become visible at certain points. While they focus on moments of breakdown, early construction when testing and redesigning is taking place represents another such moment. At these moments of breakdown/change, the essential nature of such infrastructuring becomes clear.

In Big Science facilities, both historically and today, the data management systems form an essential part of the infrastructure necessary to producing cutting-edge scientific knowledge, while remaining mostly invisible to the scientists conducting experiments. Furthermore, scientific computing (including data management) has been sidelined in scholarship about Big Science and is commonly understood as not being part of the experiment. This means that we lack understanding of the role played by data management infrastructures in the knowledge production process, and awareness of the contribution made by data management workers to the advancement of scientific knowledge.

Data as power

While data is, first and foremost, the results from an experiment on which scientists will base their results, it is also an asset in many other ways and to other audiences. For this reason, studying data at a Big Science facility is about much more than just experimental results. It is about power and politics too. In this section, I outline some of these interested parties. Many big science projects are built upon collaboration and cooperation between different research groups, and thus the sharing of data or development of data management techniques comes to play an important role in mediating the relationships between researchers and technical experts (Akrich, 1992; Beaulieu, 2001; Edwards et al, 2011; Murillo et al, 2012). However, given the size and cost of Big Science facilities, regional and international relationships related to data management are also in the spotlight in terms of financial investment, prestige and benefits.

As early as 2009, for example, the development of a dedicated data centre in Copenhagen was announced as a demonstration of the cross-border collaboration between Sweden and Denmark behind this project:

> ESS will be a world-class research facility and the governments of the two countries predict that it will have a very positive effect, both in terms of research and education in the Oresund region and for Sweden and Denmark in general. At the same time, an IT-center will be built, which will be located at Nørre Campus in Copenhagen and will provide 65 new IT-work positions. (ESS, 2009a)

The ESS is based on the premise of European collaboration, with multiple countries contributing to the costs of building this facility through a model of in-kind contributions (ESS, 2009b). Sweden and Denmark are responsible for the majority of the costs, with Sweden contributing 35 per cent and Denmark 12.5 per cent. These major contributions find concrete form in the location of the facility: the neutron facility is located in Lund, Sweden, and the DMSC in Copenhagen, Denmark. In the press release mentioned earlier, emphasis is placed on the economic benefits for the region in the form of new jobs. In this way, acquisition and analysis of data are publicly acknowledged as a key aspect of the project and constitute a highly visible way of marking Denmark's collaboration in the project, guaranteeing access and prestige for the University of Copenhagen in particular. The establishment of a new facility such as the ESS thus links national and international politics with university and research politics.

Given the enormous expense of building Big Science facilities, it is perhaps unsurprising that there is a notable demand for such facilities to provide '(demonstrable) productivity and efficiency' in order to justify public investment (Heidler and Hallonsten, 2015; see also Hallonsten, 2013). While publications produced by scientists are increasingly being used as a means of measuring the 'success' of a facility (Hallonsten, 2014), another area in which debates over who benefits can be seen is in relation to the ownership and use of data.

Formal agreements relating to sharing of and access to data are not restricted to team protocols. National and international research grants often stipulate particular requirements for access to and storage of data. Where there are commercial considerations, ownership of the data also requires careful negotiation. Access to data marks the boundaries of the Big Science community, and therefore those who are allowed to speak, to witness (pace Shapin and Schaffer, 2011).

In the case of Big Science, technologically mediated exclusion zones play an important role in determining who can take part in, or whose work is visible and validated, the cutting-edge knowledge production process at such facilities (Couldry and Mejias, 2019; Couldry and Mejias, 2020; Haraway, 1997).Technical expertise in relation to using data management also creates a boundary line between those who already have the ability to work with data management software and the many 'outsiders' dependent on the computer experts for help. Finally, it is also a question of who 'owns' the data and therefore may use it.

Negotiating access to data generates new working relationships as well as reinforcing existing bonds, creating networks of influence and dominant ways of understanding the data. Access to data is an increasingly high-profile topic both inside and outside specific scientific communities, with research grants often stipulating particular requirements for access to data. Meanwhile, open data movements demand free access to certain kinds of data for everyone to use and republish. These tensions highlight the increasing importance and visibility of data management both within Big Science itself and in popular understandings of this work.

Open data organisations demand free access to certain kinds of data for everyone to use and republish. Publicly funded research that takes place in Big Science facilities, it could be argued, is owned by the public and therefore should be made more easily accessible. This has found concrete form in organisations such as the Open Data Foundation whose aims include 'improving data and metadata accessibility and overall quality in support of research, policy making, and transparency' (Open Data Foundation). As Lucila Ohno-Machado also points out in her 2012 commentary piece in *Science Translational Medicine*: 'Today, there is little question that responsible data sharing is a necessity to advance science —and also a moral obligation' (p 1). Some large projects, for example the Human Genome Project, have already attempted to do this. However, to return to one of the challenges discussed earlier, data sets that emerge from Big Science facilities are characterised by scale and complexity; making sense of Big Science data is an activity that requires specialist skills in data management that the majority of the public do not have.

Experimental data is thus much more than just the basis for scientific results. Changes in data volumes and complexity demand material resources and technical skills, ownership of and access to data denotes membership of a prestigious scientific community that often reproduces intersecting power relations based on gender, geographical location, and disciplinary background. Meanwhile, physical location of resources related to data shows its value as a political asset for regions and nations. Data is never just data.

Structure of the book

In order to answer the question of why data management is not visible and what this matters to Big Science, I invite the reader to follow the development of the ESS as it is under construction, offering an insight into the delicate balancing act between user assumptions, technical limitations and organisational politics performed by the team leaders of the DMSC as they develop systems capable of capturing, processing, analysing and visualising vast quantities of data in real-time as experiments are performed. Chapter 2 frames this work by drawing on literature from STS and also from CDS to show what is known so far about how knowledge is produced at Big Science facilities and in what ways data may be shaped by their context of production, collection and use, respectively. This is followed by a discussion in Chapter 3 of the concept of 'invisible work' that shows how the concept is understood and used here, as well as connecting it to studies of infrastructuring and the black box. It explores possible tensions between the apparently static nature of the black box and the emphasis in infrastructuring studies on an ongoing process. Chapter 4 provides a detailed account of the fieldwork that was undertaken, including materials, methods and reflections on the role of the researcher. It includes samples of the interview questions that were used during conversations with the participants to organise the dialogue around three themes: technologies, people, organisation. It also discusses the experience of being a researcher inside the 'black box', present as the technologies are developed around me. Having set the scene and outlined the critical tools to hand, Chapters 5 through 7 each focus on one of the three themes detailed in the methods chapter. Chapter 5 dives into the technical aspects of data management, tracing the path that data take from detector to scientist. Chapter 6 turns the focus to the Group Leaders of the DMSC themselves and asks if they and their colleagues are technicians or scientists; does the work done by the DMSC fall within the borders of The Experiment or are they technical support? This leads naturally to Chapter 7 where organisational structures and politics enter the discussion more explicitly, examining how the DMSC relates to the rest of the ESS organisation. Chapter 8 draws on the empirical material presented and analysed in Chapters 5–7 in order to tackle the question of what counts as 'data'? What are the consequences for knowledge production at Big Science facilities if we situate the 'data' by highlighting the context of its production? How can we reconcile the CDS maxim that 'raw data is an oxymoron' with the requirement for 'objective' data on which to ground reliable, reproducible science? Chapter 9 is the concluding chapter of the book in which I return to the big picture of changing notions of Science and how these more critical perspectives on data and knowledge production might play a role in this.

2

A Neglected Aspect of Big Science

In this chapter, I will unpack and extend my discussion to detail the fields towards which I nodded in the Introduction. In approaching the data management technologies of the European Spallation Source (ESS) as the result of technical affordances, human experience and organisational structures, this book is primarily in conversation with two bodies of literature: Critical Data Studies (CDS) (and particularly studies of large scientific research projects) and Science and Technology Studies (STS) (particularly accounts of Big Science facilities and knowledge production therein). Conceptually, 'invisible work' will bridge the two bodies of work and connect studies of Big Science and CDS. Theoretically, it draws on the 'black box' as a way to explore data management in the context of Big Science. These concepts will be the focus of the next chapter. Here, I will introduce the scholarship on Big Science that has been produced primarily by researchers from STS. These studies are mostly organised around studies of particular facilities or scientific communities that use these facilities. My goal with this chapter is to 'set the scene' by detailing which facilities, and which aspects of facilities, have already been the subject of STS literature. In doing so, I will make clear where there is a gap in the existing scholarship around data management in Big Science. This leads me to the second major field of scholarship with which this book is in dialogue: CDS. This is a newer field than STS but one which has already engaged extensively with different aspects of data management. In both of these fields, there are important methodological and theoretical insights among the empirical studies of different facilities or different data sets on which I lean in this book. These are not necessarily always well aligned, and part of the discussion in the latter sections of this chapter will concern how to manage tensions that emerge between the different approaches.

In framing the existing literature in this way, I want to zoom in on the specificity of Big Science as a distinct context of knowledge production, one which STS scholars seem to have come closest to capturing in their accounts of different facilities. I complement this with the critical attention to context of data production that characterises CDS. The puzzle at the heart of my exploration lies here in the overlapping region between the two sets of scholarship; the production of experimental data is the raison d'être of Big Science facilities, yet strangely overlooked in the scholarship about such facilities. Meanwhile, attention to the context of data production is at the heart of CDS, yet that field of scholarship has not yet paid much attention to the distinctive context that is Big Science facilities.

Big Science: some back story

In the following two sections, I will provide first an overview of the scholarship produced to date about Big Science and then specifically about the ESS as an example of Big Science. I will not be dealing with the enormous body of literature that has already been produced that focuses on aspects such as the development of instruments at Big Science facilities or the neutron source, and which is typically published in journals such as *Nuclear Physics B*, *Journal of High Energy Physics*, *Journal of Neutron Research* or similar publications. Instead, my focus here is on how Big Science has been discussed within the field of STS, an area of scholarship which brings together expertise from across the social sciences and humanities to examine *how* knowledge is produced and science is 'done', with a particular attention to the effect of the surrounding sociocultural context. Work conducted under the rubric of STS typically (although not always) involves qualitative research comprising, for example, ethnographic studies of daily laboratory practices or discourse analysis of science policy documents. It pays attention to how stories and imaginaries about science and technology are interwoven with the actual *doing* of it. As a multidisciplinary field it brings together scholars from a wide range of disciplines, including but not limited to sociology, history, anthropology, library and information studies, and in my case, gender studies.

Studies of Big Science facilities constitute a relatively small area of interest within STS, exemplifying as they do a particular context of knowledge production. To date, there are a number of studies that provide the history of well-known facilities, giving the reader a detailed look inside the development of prestigious laboratories such as CERN (Hermann et al, 1987; Krige, 1996), the Lawrence Berkeley National Laboratory (Heilbron and Seidel, 1989), the European Synchrotron Radiation Facility (Cramer, 2017, 2020; Simoulin, 2017), the European Southern Observatory (Blaauw, 1988) and the Brookhaven National Laboratory (Crease, 1999), among others. The appeal of such volumes is clear; they give a glimpse into places where discoveries that change the world have taken place, and which continue to drive innovation. Reading these volumes highlights the ways in which national research politics, personal agendas and advances in scientific understandings were entangled. They also engage in a broader discussion about what constitutes 'big science'. Unsurprisingly given the longevity of these facilities (the Berkeley Lab was founded in 1931, and CERN in 1954), several of these accounts run to multiple volumes. Other accounts explore the specific conditions which unfolded to make the founding of specific labs possible (for example, Hoddeson et al, 2019 discuss Fermilab) or not (Riordan et al, 2019, for example, discuss the failure of the SuperConducting SuperCollider project). In addition to these book-length studies, there are numerous articles exploring particular aspects of individual facilities. Finally, and of high relevance to this book, are those works which have highlighted particular sociocultural configurations within those working at Big Science facilities, such as the role of technicians (Doing, 2009) or how the practices involved in learning how to be a physicist differ across national cultures (Traweek, 2009) or time (Galison, 1997).

Doing's study of the Cornell Synchrotron, *Velvet Revolution at the Synchrotron: Biology, Physics and Change in Science* (2009) is particularly useful here for three reasons. First, his focus is on the machines used at a Big Science facility not as static entities but rather as dynamic, changing parts of a broader scientific assemblage that encompasses humans, technologies and organisations. There is therefore a great deal of movement and change in his account, and he connects changing research agendas on a national level with institutional politics and everyday practices in a way that produces a very user-friendly account of a complex process.

Second, he focuses on the role of those operating the machines, rather than the scientists themselves. Using this lens, I am interested in the role played by the data 'operators' of the ESS, the Data Management and Software Centre (DMSC), and the technologies they are designing, developing and deploying in the pursuit of science. His discussions of the different roles in the lab in relation to the practice of doing Science were particularly relevant to my discussion in Chapter 6 about my participants' backgrounds.

Finally, he incorporates a high level of reflexivity into his own account that contributes to broader scholarly discussions about the role of the technician. In *Velvet Revolution at the Synchrotron* Doing narrates his own developmental trajectory working initially as an 'x-ray laboratory operator' and later as 'assistant operations manager'. During this time, he simultaneously pursued a PhD in STS. He therefore occupied the double position of insider and outsider, participant and observer. The resulting study, although focused on x-ray machinery rather than data technologies, is highly relevant to this book as a study of 'whether the product of scientific practice, the very content of science, is contingent on that practice' (Doing, 2009, 23). In an early chapter of his book, Doing summarises a number of key 'laboratory studies' from STS which – through detailed observations of 'science-in-action' have demonstrated the contingency of science as we know it. Doing's primary critique of these studies is that they often stop short of considering the role played by the machines used by scientists to calibrate, verify and test their hypotheses. In *Velvet Revolution* the reader is allowed to get up-close with the nuts and bolts of the machine Doing is operating in an effort to understand how such machines are a contingent part of the experiment, replete with limitations and affordances, whose operation intersects with human expectations and epistemological positions.

For the most part, the volumes listed earlier deal with those who have historically been the most common users of these facilities, the high-energy physics community, and the development of particular facilities to serve this user group. As noted in the Introduction, however, more recent scholarship has reflected on how changing user groups and practices are challenging what counts as 'Big Science' (Cramer and Hallonsten, 2020). The question of just what is meant by the moniker 'Big Science' is up for debate, as the user group for facilities like the ESS becomes more multidisciplinary, other kinds of experimental setup demand just as costly equipment, and enormous, international, dispersed collaborations such as those working with climate science or biology involve many more people (Edwards, 2010; Vermeulen, 2016). Cramer et al writing in the opening chapter of *Big Science and Research Infrastructures in Europe,* for example, note that 'On the basis of the current use of the term Research Infrastructures and the historical and current use of the term Big Science, it can be established that the categories of things that the two describe are very wide and partially overlapping' (2020, 1).

The authors' goal in this anthology seems to be to expand understandings of 'Big Science' predominantly by placing it in conversation with the concept of 'Research Infrastructures'. Similar moves to reframe and update the notion of 'Big Science' can be seen in the suggestion of terms such as 'Big Science Transformed' or 'New Big Science' (Crease and Westfall, 2016; Hallonsten, 2016; Rekers and Sandell, 2016). But what makes these facilities different?

> Firstly, that they are so large and expensive that they are beyond the scope of most regional and national budgets, and instead require several countries to collaborate. Secondly, they are expanding in terms of their multidisciplinarity. In addition to scientists from physics as well as chemistry and the life sciences, new users come from disciplines such as archaeology, geology and medicine. Thirdly, they are expansive in their ambitions to contribute to society, and are thus shifting the way promises are made. (Rekers and Sandell, 2016, 8)

> The evasiveness and near analytical uselessness of the term and concept 'Big Science' stems partly from its entering into popular culture (Capshew & Rader 1992:4; Hevly 1992: 355) and partly from the

> inability of pure *size* to ever explain *content* (Westfall 2003). (Hallonsten, 2012, 81)

Both of the previously discussed anthologies emphasise the various ways in which political and social changes have impacted how research at Big Science facilities is organised and categorised. Curiously, the increase in volumes and complexity of data produced at such facilities is rarely discussed.

In essence, what I have tried to show earlier is that the STS canon of books about Big Science has historically focused on the stories of specific physical experimental facilities that emerged from Cold War efforts to drive scientific advances. Inevitably, the definition of what 'counts' as Big Science has changed and stretched to reflect different social, political and research conditions since then, with no clear consensus on what such large-scale efforts should be called or how to define them. Therefore, understanding these shifts is necessary to understanding something of the disciplinary foundations upon which experimental technologies and techniques in Big Science facilities like the ESS have been grounded, and which are reflected in the academic pedigrees of my participants.

Big Science at the ESS

The discussions summarised in the previous section concerning the changing role/status of Big Science are highly relevant to the case of the ESS, as the volumes that deal specifically with this facility make clear. Although the ESS has been under discussion for more than 20 years, it is only in the last approximately ten years that STS volumes on the facility have started to appear. This reflects the leaps and pauses in development which saw the facility only become an organisational unit in 2009. By 2012, when for example the various authors of *Legitimizing ESS* were studying the facility, there were only 100 employees, and it was still unclear if the network of European partners would come through with the full financing. The primary volumes that have examined the ESS to date being: *New Big Science in Focus: Perspectives on ESS and MAX IV* (Rekers and Sandell, 2016), *Legitimizing ESS: Big Science as Collaboration Across Boundaries* (Kaiserfeld and O'Dell, 2013), and *In Pursuit of a Promise: Perspectives on the Political Process to Establish the European Spallation Source (ESS) in Lund, Sweden* (Hallonsten, 2012). More recently there have been doctoral dissertations, such as 'Accounting the Future: An Ethnography of the European Spallation Source' (Dimitrievski, 2019). These studies take different perspectives on the ESS, spanning history, politics, environment and more. As their titles suggest, they each have a slightly different departure point.

The first to be published, *In Pursuit of a Promise: Perspectives on the Political Process to Establish the European Spallation Source (ESS) in Lund, Sweden,* has as its aim to reveal the 'real story' behind what happened when a decision was made about the location of the ESS. Complementary to this goal, was a perceived gap in the social sciences literature about the ESS and a need to increase public interest in the project: 'Social scientists, like the authors represented in this volume, certainly have a responsibility to critically analyze issues of this sort, not only to satisfy academic interest, but also as a public service, and in the case of the ESS this seems rather urgent' (Hallonsten, 2012, 13). As the title of this volume suggests, its focus is on the policy perspective, and on lifting up the specificities of the Swedish context.

Next to be published was the 2013 volume *Legitimizing ESS: Big Science as Collaboration Across Boundaries*, in which the authors aim to 'explain the complexity of the cultural, social, and political processes that determine Big Science' (p 8) as well as 'to study the key process leading up to the realisation of the ESS' (p 9). In order to achieve this, the book sells its unique contribution as being on the diverse disciplinary perspectives provided by the authors. This means that the book promises chapters as diverse as 'The ESS in the local news media', 'The ESS and the geography of innovation' and 'Designing for the future: Scientific instruments as technical objects in experimental systems', among others.

Finally, *New Big Science in Focus: Perspectives on ESS and MAX IV* was published in 2016, as the result of a specifically interdisciplinary project around the ESS, which sees many of the chapters in this volume in dialogue with one another. This volume has perhaps the most forward-looking standpoint: 'Our aim has been to identify emerging areas of interest in new big science, to present initial findings from our empirical investigations, and to indicate interesting themes for the future' (2016, 9). In keeping with its attention to how Big Science is changing and becoming 'new big science', this volume opens up new topics for investigation within Big Science, which have not been addressed in earlier volumes, namely sustainability, data management and legal questions.

It is notable that unlike the volumes mentioned in the previous section, those detailing the ESS's development are all – with the exception of the doctoral dissertation – coauthored volumes in which authors from a wide range of disciplines (but still within the social sciences and humanities) bring their specific expertise to bear on the topic. This collaborative effort allows the complexity of such a project to emerge more clearly, as different authors present quite different perspectives. It is also worth noting that the editors of all three volumes were located at/closely connected to Lund University, and thus perhaps more aware of the local discussions that had surrounded the development of the facility. Furthermore, the same authors appear across all three volumes giving some sense of the relatively small group of scholars who have studied the ESS to date. Finally, all three volumes engage with the tricky question of how to understand Big Science, and how the ESS relates to previous 'Big Science' facilities.

Comparing the STS literature on Big Science with those volumes that have been produced about the ESS it is possible to see similarities in themes that emerge, but also new emerging areas such as two chapters related to data at the ESS. Overall, however, the literature on Big Science facilities as sites of knowledge production to date lacks an in depth discussion of data and data management, despite the importance of data collection as the basis for experimental results. In the following section, I have collected the few scholarly contributions on data in Big Science that I have been able to find.

Data in Big Science

What little scholarship there is available concerning data in Big Science appears as the focus of a handful of papers, as well as part of the discussion in larger volumes. Perhaps most notable among these is Peter Galison's *Image and Logic: A Material Culture of Microphysics*, in which data appear in relation to changing experimental technologies and techniques. For example, in Chapter 6, titled 'The Electronic Image: Iconoclasm and the New Icons,' he discusses the role of the visual image in experimentation as the era of bubble chamber experiments came to a close. He argues that the shift from visually oriented experimental approaches (in which experimenters waited patiently to capture a 'perfect specimen of an *X* or *Y* decay' (1997, 434)) towards an approach where experimenters used new instruments to 'manipulate the world at the finest level' (Galison, 1997, 434) also marked a shift in how data was understood: 'Experimentation in the aniconic tradition of the logic experimenter was getting results as they occurred, not waiting weeks for a photolab to return ten thousand pictures, and obtaining results by way of a scanning squadron' (1997, 434).

In this book, Galison accounts the ways in which experimental practices, and more precisely the machines used in such experiments and the notion of the experimenter themselves, changed in 'microphysics' (what later came to be known as particle physics). In doing so he studied one of the primary user groups of Big Science facilities, tracking the shifting relations between theory and data over many decades. For example, recounting the charm experiments conducted in the 1970s when unexpected results challenged pre-existing ideas, Galison engages with the tension between antipositivist and positivist discussions about experiment-theory relations: 'On the old positivist view, data were hard and theory ephemeral. On the view espoused by the antipositivists, data are "tunable" and theory powerful and controlling. Neither view seems adequate' (1997, 543).

Galison goes on to state that 'data are always already interpreted' (1997, 543), but he does not mean interpreted by a theory. Instead, he brings us back to the machines themselves, stepping us back through an experimental procedure in search of what might be understood as 'raw' data. Each time he considers how a machine or a process adjusts, discards or interprets data he comes up empty-handed in his quest for 'raw'. His narratives reveal how each machine has been designed or programmed in a particular way ('the data are still not in that Edenic state', as he puts it so eloquently (1997, 543)). Finally, he concludes that 'there are no original, pure, and unblemished data. Instead, there are judgements, some embodied in the hardwired machinery, some delicately encoded into the software' (1997, 543–4). Galison resists, however, the poststructuralist desire to deconstruct the truth of the data into oblivion, instead reminding us of the long tradition of instrument design and testing, showing how each innovation builds on the previous generation of tried-and-tested machines to constitute a 'series of certifications'. Galison's account emphasises how this accumulation of expertise and experience provides the grounds for reliable results, while acknowledging how material limitations of technologies may shape the collection and management of data. Among these technologies, the key role played by computing is made clear from the opening pages:

> In 1964, some of the world's leading experimental physicists gathered in Karlsruhe, West Germany, to discuss the radical changes then underway in their profession. … Where physicists once collected data in notebooks and analyzed them with slide rules, the computer had now taken over much of this work – storing, processing, even analyzing information and delivering it in publishable graphic form. … Throughout the laboratory, the relations among computer programmers, experimenters, instrument makers, and engineers were utterly in flux. (Galison, 1997, 1)

The increasing role of computing power in Big Science experiments is also in focus in a pair of papers by Robert W. Seidel, who recounts the history of the introduction of scientific computing during the 1960s, 1970s and 1980s. Seidel pays attention to the organisational dynamics that controlled funding and leadership of scientific computing, painting a picture of ad-hoc deployment. While his focus is on computing, he makes it clear how this is integral to discussions about change to data management; 'the role of computing in data acquisition is of singular importance in modern HEP collaborations' (Seidel, 2008, 481).

In Seidel's article, 'From factory to farm: Dissemination of computing in high-energy physics' (2008), the focus is primarily on the impact that computing developments made to the experience of the experimenter.

In other words, while the technologies required to handle ever larger volumes of data are visible and explicit in his account, the selection of such machines, their installation, maintenance, and development by another group of experts is missing. He writes, for example, about the advances in instruments and detectors that took place during the 1980s: 'This inspired a new departure: event-oriented data analysis, which required greater computing capacity and capability, as well as a division of computer labor that brought the physicist greater insight and control into the physics of an experiment' (2008, 481).

In his fascinating account, Seidel offers a number of clues as to how local conditions shaped the development of data management tools and techniques. He describes how increasing numbers of images produced by bubble chambers (which were deciphered by human scanners) ultimately led to a desire for automated data analysis. The limitations of human processing power that spurred on computing developments also resulted in an outsourcing of data analysis training: 'While a photograph was simple enough to allow technicians to identify significant events with a little training, automated data analysis required much more rigorous programming, which was usually left to the users' computer facilities in universities or other independent research organisations' (Seidel 2008, 485).

Thus, technical advances in detection spurred on technical advances in computing, resulting in organisational changes and a different team being involved in the support around experimentation. Similarly, he notes the personal reluctance of the lab's director in tandem with knock-on delays from construction of the accelerator itself as leading to a rather late acquisition of computers at Fermilab. In '"Crunching numbers" computers and physical research in the AEC laboratories' (1998) Seidel goes further back in time to the 1940s and the early development of computers at sites such as Los Alamos. His account shows how close collaboration between the Atomic Energy Commission, universities and scientific community resulted in important breakthroughs. Here both machines and people are in focus, with the development of some machines reading like a roll call of eminent computer experts. Seidel's account here shows how expertise travelled between different sites and companies, resulting in a national dialogue conducted through the movement of human bodies, as knowledge was handed down and shared. While data is not in focus in this piece, Seidel's historical account makes clear how inherited expertise passed through a community shapes what the next generation develops. In this way he highlights how cutting-edge technical developments often occur as the results of informal networks, an important sociocultural context to note when also considering the development of data management computing.

In summary, these historical accounts lift up different aspects of data management and scientific computing. Galison discusses the machines in a sociocultural vacuum, in which interpretation of the data takes place through technical functionality, with little relation to those who built the machines, or the organisations that funded them. Seidel's narrative, in contrast, lifts up wider structures such as communities of experts, institutional politics and research dynamics as shaping the development of Big Science.

Returning to the present, there are a handful of important papers about contemporary data management worthy of note. 'Data in the Making: Temporal Aspects in the Construction of Research Data' by Jutta Haider and Sara Kjellberg (2016) focuses on how data will be managed at MAX IV and ESS. In their contribution to the *New Big Science in Focus* anthology, they show how different groups at ESS and MAX IV understand 'data' differently. To do this, they focus on temporality and particularly the question 'when are data?' as a way to explore these shifting understandings across different phases and roles within research: 'The making of data does not refer here solely to the data produced during an experiment or an observation, but rather to how they are made possible by setting up and planning for the production, storage, and use of data, and even the limitations, strategic roles, and other effects' (2016, 143). They frame their study in the broader landscape of open data and changing research policies and are perhaps less focused on teasing out what may be distinctive in the Big Science context or the technical details around data management.

Koray Karaca's work provides detailed insight into contemporary technical measures in place at CERN and the Large Hadron Collider for data management. In 'What Data Get To Travel in High Energy Physics?' Karaca acknowledges that HEP has become 'progressively data intensive over the past 60 years' (2020, 45):

> I will focus on the initial data journey in the ATLAS experiment that links the production of collision events at the LHC to the stage of data acquisition where *usable* data are constructed out of collision *events* detected by the LHC, prior to the stage of data analysis and modelling. (2020, 46)

Through an extremely detailed analysis of the data journey undertaken Karaca is able to show how technological limitations impact each stage of the early data journey: 'the full description of the event is not yet known, and as a result, the level-1 event selection is performed without *full granularity*, that is, without the availability of data from all the channels of the individual detectors' (Karaca 2020, 51). Karaca is at pains to explain how the automated data management system has been carefully designed to capture particular kinds of events based on existing scientific knowledge about the kinds of particles produced during experiments. However, he also shows how the process of transforming collision events into usable data requires particular algorithms that in essence reconstruct the full story of the collision from the fragments collected:

> The first part of the local journey is a construction process in the sense that event fragments are assembled by the level-1 and level-2 triggers into *full* events. This part of the local journey is at the same time a selection process, because both events and event fragments that do not satisfy the selection criteria are filtered out and discarded from further consideration. (2020, 54)

Karaca is also at pains to draw the reader's attention to the material limitations of the technology, providing a helpful example of what these might be in the case of the LHC:

> The technical limitations in terms of data storage capacity and data process time make it necessary to apply data selection criteria to collisions events themselves in *real time*, i.e., during the course of particle collisions at the collider (ATLAS Collaboration 2012). Moreover, due to the aforementioned technological limitations, only a minute fraction of the interesting events could be selected for further evaluation at the stage of data analysis. (2020, 48–9)

The material affordances and limitations of technologies might include processing power, bandwidth or storage space, all of which might affect how data are collected or managed. As such, Karaca's account is a useful complement to my study, as it provides an account of an equivalent part of the data journey at a facility that is running, rather than under construction. It also doesn't hesitate to dig into the technical details. However, perhaps because of this attention to technicality, or because it is an automated system, the reader gains little sense of the surrounding sociocultural or organisational context that shaped development of the algorithms or determined what the funding was for data capacity.

Also engaged in studying the LHC is Antonis Antoniou, who examines the process of constructing a 'data model' during HEP experiments on Lepton Flavour Universality. Antoniou's work draws on discussions within CDS (Leonelli, 2015, 2019; Bokulich, 2020, 2021) to engage with philosophy of science discussions concerning the relationship between data and phenomena (McAllister, 1997; Glymou, 2000; Harris, 2003). This is grounded in Patrick Suppes' conjecture that 'theoretical hypotheses are not directly confronted with the raw unprocessed data from experiments, rather, they are only confronted with *models of data*' (as it appears in Antoniou, 2021, 101; see also

Suppes, 1966). In walking the reader through the data processing procedure during experimentation Antoniou is able to demonstrate that:

> the first data collected at the early stages of the experiment, which can be characterised as the raw data of the experiment, are useless as they are for the comparison between theory and experimental results, since they necessarily need to undergo a process of refinement in order to be transformed into a language that is comparable to theory. (2021, 101)

This observation contains interesting echoes with the historical accounts of data in Big Science, continuing as it highlights once again the theme of increasing complexity and user-unfriendliness of data that is familiar from Galison's account. In both we see how, as experiments and technologies develop, an increasing amount of processing/support appears to be necessary in order to render the data into something which human users can parse.

Although a relatively small group of scholars, the work outlined previously provides a remarkable breadth in terms of contextualising data management in Big Science, drawing attention to the materiality of the machines, the political and organisational dynamics, epistemological framings and software limitations. The more recent scholarship is indebted to varying degrees to the field commonly referred to as CDS, the fundamental premises of which I discuss in the following section as an important inspiration for my own study.

Critical Data Studies: basic concepts

> Just as big science was to reveal the secrets of the universe, big data is expected to reveal the buried treasures in the bit stream of life. (Borgman, 2015, 3)

> Does the computer not only observe, but also create, new data, as accelerators and lasers do? (Seidel, 1998, 32)

Christine Borgman opens her 2015 volume *Big Data, Little Data, No Data* by invoking Derek de Solla Price and his comparison of 'big science' and 'little science' as a metaphor for the promise and hopes pinned on both big data and big science to offer solutions to society's 'big' problems. While data management may have received relatively little attention in studies specifically focused on Big Science, it has been well discussed within what has been termed 'Critical Data Studies' (Dalton and Thatcher, 2014; Kitchin and Lauriault, 2014). The impact of digitised information management on knowledge production has been explored within studies of infrastructure (Edwards et al, 2011; Bietz and Lee, 2012) and as part of the literature on invisible work (Ehrlich and Cash, 1999; Star and Strauss, 1999). Critical studies of big data by communication and digital media scholars such as Borgman (2015), Gitelman (2013), Hine (2006) and Kitchin (2014) draw attention to the ways in which decisions about data management and virtual infrastructure constitute a form of knowledge production. These often draw upon STS concepts such as 'situated knowledges' to stress that 'how data are conceived, measured and employed actively frames their nature' (Kitchin and Lauriault, 2014).

Borgman's work, for example, focuses on temporality in relation to data. By asking *when* are data, Borgman highlights the changing status of data in the case of reuse of data, data generated from simulations and metadata, and the consequences of processing data. More recently this has been developed by Sabina Leonelli who thinks with the notion of the 'data journey':

> data that go through journeys are rarely unaffected. Travel can affect their format (e.g. from analog to digital), their appearance (when they are visualised through specific modelling programs), and their significance (when they change labels, for instance when entering a database). Travel also typically introduces errors, such as those caused by technical glitches plaguing the transfer of data from one type of software to another, power cuts or lack of storage space when moving data to a new server, or typos made when manually inserting variables into a database. (Leonelli, 2019, 41)

In the case of the ESS, the data management journey involves data being collected from detectors surrounding the experimental instruments and fed into a distributed streaming platform. From there it will be distributed in a least two ways – one to a data writing/archive facility and one to a real-time analysis program that will allow users to adjust ongoing experiments. This set-up is by no means standard and every facility arranges this differently. At each stage, the data acquires various kinds of metadata and is unpacked/repacked/tagged thereby acquiring additional layers of meaning and information.

> The metaphor of the 'journey' is powerful because, just like many human journeys, data journeys are enabled by infrastructures and social agency to various degrees and are not always, or even frequently, smooth. A useful way to think through the significance of adopting this metaphor is to consider what it can mean for journeys to be successful. (Leonelli and Tempini, 2020, 9–10)

Working with the notion of a journey allows me to pay attention to the impact of the diverse technologies on the production and collection of experimental data, as well as the points at which different parts of the journey stop, start and join up … and the inevitable 'lost luggage', cultural gaps, time lag and other discomforts associated with travelling. It is realistic about such difficulties and thus poses the question of what level of tolerance is built into a system such that a journey is considered successful. It thus brings together a number of theoretical and analytical tools in order to answer the question: what happens 'behind the scenes' of a Big Science facility to smooth the seams in data capture? In studies of other kinds of 'Big Science' the idea of a 'data journey' has been useful for tracing the movements of data across the seams of large, distributed networks of collaborators in relation to, for example, climate data (Edwards, 2010).

Examining the journey which data take early on in the experimental process provides not only insight into the technical affordances or limitations of the hardware and software, it also provides a way in to examining the assumptions about its use later in the process: 'Data are not pure or natural objects with an essence of their own. They exist in a context, taking on meaning from that context and from the perspective of the beholder. The degree to which those contexts and meanings can be represented influences the transferability of data' (Borgman, 2015, 18).

It is here – even before the data reaches the users at the ESS – that reflecting on user assumptions becomes pertinent. Identifying who is the assumed user of the data can reveal the ways in which the data is shaped to their purposes as it is collected and processed, and simultaneously, which uses and users may encounter difficulties transferring or re-processing that data, for example due to different disciplinary or institutional guidelines for metadata or archiving protocols (Haider and Kjellberg, 2016).

The use of the journey metaphor brings attention to not only time but place – not just *when* but also *where* is data at the ESS? These different kinds of questions have opened up a critical space for considering the context in which data is conceived and produced, as well as considering the consequences of such questions. Gitelman and others have drawn attention to a prevailing attitude that 'data are apparently before the fact' leading to 'starting with data often leads to an unnoticed assumption that data are transparent, that information is self-evident, the fundamental stuff of truth itself. If we're not careful, in other words, our zeal for more and more data can become a faith in their neutrality and autonomy, their objectivity' (Gitelman, 2013, 2–3).

As exemplified by Borgman's and Gitelman's work, underpinning many of these volumes is a call to interrogate how we understand 'data' itself: 'Data need to be imagined *as* data to exist and function as such, and the imagination of data entails an interpretive base' (Gitelman, 2013, 3). They draw attention to the ways in which epistemological paradigms define 'data' differently but also that these definitions are entangled with specific disciplinary research practices and methodologies concerning how 'valid' data should be produced, processed and stored, and thus how 'objectivity' is achieved. CDS scholarship therefore 'situates' data as the product of a specific context and enquires into the consequences of such 'situatedness' for the knowledge that is then derived from the data. This scholarship illustrates Bowker's 2008 provocation that 'raw data is both an oxymoron and a bad idea' (see also Bowker and Star, 2000), by highlighting how specific disciplinary and epistemological contexts influence decisions and practices to 'cook' data in different ways.

While the idea of 'cooking' data is popularly associated with fraudulent behaviour,[1] in the context of CDS scholarship 'cooking' acts as a metaphor or prompt to enquire into how software, hardware, organisational structures, technical expertise, epistemological expectations or methodological best-practice might affect what data is collected and the subsequent processing. In my work it occupies a somewhat ambivalent position, hence the fact that it always appears in scare quotes. When talking to peers in STS or CDS 'cooking' is a useful shorthand for connecting context with data. In the context of Big Science, however, the idea that data might be 'cooked' troubles the reliability of the experiment. For those engaged in the hard work of producing experimental data in Big Science, 'cooking' might be perceived as an accusation that data have been altered in ways that would disturb their validity, and was a term I tried to avoid using with my participants. This double use (analytical tool vs actor category) is present throughout my study but I return to discuss it in depth in Chapter 8.

Conclusions

Processing of data by various kinds of software, for example, aims to 'return meaningful and scientifically valid data back to its users' (ESS, 2015). However, it also calls into question the status of the information collected from an experiment *before and during* the processing of this information that takes place on its way to appearing on a scientist's screen. Does it 'count' as data before the 'magic moment' (Borgman, 2015) that a scientist first sees the results? At what point do I and my participants consider that a 'cooking' process starts (if at all), and what might be learned from these different understandings of 'raw' data? Given the paucity of discussion about data in the histories of Big Science to date, I propose to bring together scholarship from CDS and STS in order to address this lack and suggest how data management constitutes a form of 'situated' knowledge production (Haraway, 1991) in New Big Science (NBS).[2] Or to borrow from Dalton and Thatcher,

[1] For example, 'cooking the books' is a metaphor for committing financial fraud.

[2] A similar move – between media studies and STS – has been proposed by Lievrouw (2014) in order to address the lack of attention to materiality within much media studies scholarship.

for what purpose is data created at the ESS: 'The data of "big data" can take many forms for many purposes: from the massive streams generated by the Large Hadron Collider to the global corpus of tweets. In each case, the data's format and content have been shaped and created for a purpose' (Dalton and Thatcher, 2014).

In line with the CDS scholars I mentioned earlier, this book is not only a story about how the DMSC developed over three years. It is also a story about 'raw' data and objectivity, and how my own view on 'raw' changed. At stake for me is a commitment to the critical view of the world encapsulated by Geoffrey Bowker's maxim: 'raw data is an oxymoron'. Identifying both as a feminist STS scholar fascinated by the promise of cutting-edge science and a feminist CDS scholar sceptical about the notion of 'raw data', the 'truth' I need to reveal to uphold my scholarly credentials is the way in which data journeys at the ESS will be shaped by the technologies, people and organisation surrounding them. In contrast, at stake for my participants and the future users of the ESS is the promise of raw data upon which experimental results depend; any hint of 'cooking' would invalidate the experiment. So, I'm going to tell two stories – or at least try to tell two stories.

3

In/Visibility

In the preceding chapter I outlined how different accounts of knowledge production at Big Science facilities have highlighted different aspects of the work involved. I argued that many of these volumes can be read as representative of their times both in terms of what Big Science has looked like historically, but also in terms of how Science and Technology Studies (STS) has conceptualised/approached studying Big Science. I also identified an ongoing discussion about what is meant by the term 'Big Science' and how the European Spallation Source (ESS) has been framed as exemplifying a 'New Big Science'. Following the argument advanced in *New Big Science in Focus: Perspectives on ESS and MAX IV* (Rekers and Sandell, 2016) the ESS fulfils many of the criteria associated with this new variant, namely important changes to the user group, the way the facility is understood in relation to society, and in terms of the kinds of technical support required for users to conduct experiments, most particularly in terms of data management. Two chapters in this aforementioned anthology focus on data, reflecting not only the increasingly complex legal terrain around data but also a growing critical scholarly engagement with the epistemological and methodological foundations for what 'counts' as data in these contexts. Chapters such as these, together with a small handful of journal articles, currently represent the total scholarly work concerning data management in Big Science. The low visibility of such scholarly perspectives is perhaps surprising given the emergence and growing popularity of Critical Data Studies (CDS) in recent years.

A recurring theme in the CDS scholarship is the way in which data may often be considered as neutral or transparent, with the technologies or processes through which they travel rendered somehow invisible. This lack of visibility related to data management work is also I argue a feature of the STS literature on Big Science, as well as within Big Science facilities themselves. This chapter engages with the question of *why* it might be important to make data management work in Big Science more visible, and thus lays the theoretical foundations for the following chapter on methods where I consider *how* to make this work more visible. To help me achieve this, I draw on two concepts familiar to many of those within STS: invisible work and the black box. Both are terms intimately engaged with questions of visibility, although they come with slightly different inheritances in terms of how they have been used. In this book, I am inspired by both and draw on them to support my explorations in slightly different ways. In this chapter I introduce the concepts, explore how they have been used in the literature to date, and finally consider some possible limitations of them.

With the title of this book (*Behind the Science*) I am playing around with the idea that there might be a 'front stage' and a 'back stage' in Big Science. 'Front stage' is the 'main show' of the experiment, which tends to be the one accompanied by all the grandiose rhetoric (Dimitrievski, 2019). This is also the view represented in the diagrams that appear on the ESS website and which I discussed in the Introduction (in which the DMSC appears as a smaller image dwarfed by the experimental facility organised around the source). Then there is 'back stage' where a lot of other work is done that is vital for the smooth running and successful completion of the experiments but which doesn't often get mentioned. Data management, as I argued in the previous chapter, potentially qualifies as a back stage activity. 'Behind the Science' as a metaphor is meant to conjure up questions about visibility and obscurity, but also about what 'counts' as Science in this context. The curtain between front and back stage represents a boundary line between categories such as scientist and technician, as well as between support (infra)structures and primary work. Given the changing conditions for Big Science, and particularly the ways these are manifesting in the ESS, I am interested in exploring the question of whether the curtain at the ESS is fluttering, ruffling, giving the audience a glimpse of work that has previously been going on back stage. In trying to work through these questions, I have sought support in existing, tried-and-tested concepts from STS that aim, in different ways, to make visible the back stage work carried out by people and machines. The first of these is the black box.

The black box

In *Science in Action: How to Follow Scientists and Engineers Through Society*, Bruno Latour suggests that technologies or facts (what I am going to refer to as 'artefacts' going forward) can become 'black boxes', meaning self-contained, opaque entities apparently without history or controversy (1987). The metaphor of the black box emerged initially from cybernetics (Petrick, 2020) during the 1940s as a way to demarcate part of a process where input and output were important, but where there was less concern or need to understand the processing in-between. Following Latour's use of it to unpack the technoscientific controversies around the double helix and the Eclipse minicomputer, it has been widely adopted within STS as well as other neighbouring fields as a way to describe artefacts that may appear neutral, objective or somehow obvious, but which on closer inspection contain a complex history of claims, politics and positions. The more user-friendly or simple the artefact appears to be, the harder it is to examine the history of its development and use, thereby rendering much of the work done to produce it invisible. The metaphor of the black box also points to the way in which artefacts may achieve the appearance of stability (meaning consensus has been achieved) and prompts questions as to under what conditions an appearance of stability is desired and by whom.

In order to situate the artefact within a context, and to explore the assumptions that have been 'baked in' it is necessary to open the black box. The conditions under which a black box may be opened and its contents explored have also been widely discussed in the STS literature. This can be achieved through various means, although Latour's approach favoured an historical investigation: 'Instead of black boxing the technical aspects of science and then looking for social influences and biases, we realised … how much easier it was to be there before the box closes and becomes all black' (1987, 21).

As noted by Shindell in his helpful overview of the black box (2020), an historical reconstruction is but one way in which STS researchers have approached opening the black box. Others, for example, have focused on moments of instability when the existing paradigm appears to be under attack (see Marcheselli, 2020 for a striking example) or when changing actors create instabilities (Latour, 1999; Gehl, 2016). Another way to open up this box would also be to follow a journey through the box, for example how information moves through a system (Harrison, 2015).

The black box is a useful metaphor when studying data management at the ESS, representing as it does a part of the experimental process of which most users are unaware, being primarily concerned with input (experimental sample, instrumental set-up and so on) and output (numbers and timings of detector events). In what follows I will be exploring the complex set-up that sits between this particular input and output. In doing so, I had the advantage to be there while development of this set-up was taking place. In Latourian terms, I was able to be there 'before the box closes', as the facility was under active construction during my period of fieldwork.

That said, the metaphor of the black box is not without its critics, most notably Langdon Winner's 1993 piece, 'Upon Opening the Black Box and Finding It Empty: Social Constructivism and the Philosophy of Technology'. In this he identifies a number of potential shortcomings associated with the social constructivism approach being deployed by STS scholars at the time of his writing, and exemplified by the black box. These shortcomings include too great attention to the context of development resulting in a lack of attention to the consequences of technoscientific innovations (what might simply be called the 'so what?' provocation). The second shortcoming is related to the first, in that he draws attention to a lack of reflection on who and what are considered to be important stakeholders in the development of an artefact, and were overlooked or downplayed in its development: 'although the social constructivists have opened the black box and shown a colorful array of social actors, processes, and images therein, the box they reveal is still a remarkably hollow one' (Winner, 1993, 374–5).

Winner's article engages primarily with a now well-known circle of STS scholars including, but not limited to, Bruno Latour, Michel Callon, Steven Woolgar and Trevor Pinch. Unsurprisingly the article produced a slew of responses to Winner's critiques, some more attuned to power than others. In the intervening years since *Science in Action* was published, the trope of the black box has also been widely adopted and adapted beyond STS, appearing in fields particularly relevant to this book such as CDS, infrastructure studies and platform studies.

Of particular note are feminist uses of the black box, which explicitly engage with the themes of power and benefit (Lykke, 2016; Kirtz, 2018; Smyth et al, 2020) pioneering a critical attention to context combined with a more reflective stance that limits the knowledge claims being made. The attempt to pay attention simultaneously to different aspects of context and the consequences of them on technological development has also resulted in what might be perceived as revisions/responses to the black box. A better known example of this might be Donna Haraway's 'imploded knot', an approach that seeks to disentangle the co-constitutive nature of an artefact by tracing the technical, discursive, material and sociocultural strands which feed into it. This knot embraces the messiness and dynamism of artefacts as well as the consequences of moments where artefacts solidify (1997) for bodies and lives in its environs.

The black box is one tool (of many) that emerged from a shift in Science Studies towards laboratory studies that started in the 1980s. As others have argued far more eloquently, this shift marked important changes not only in how knowledge production was understood as the outcome of a highly specific context replete with people, politics, technologies and various organisms (Doing, 2008; Garforth, 2012; Wylie, 2020). It also changed the social sciences themselves that were engaged in studying the spaces:

> In positioning the laboratory as a privileged actor in scientific knowledge-making, laboratory studies also made it central to social studies of science. An 'exemplary site' where experimental work is on show and black boxes remain open, the lab has certainly been

> methodologically convenient (Sismondo 2004, 86). But it has also been epistemologically crucial. In seminal studies, the laboratory is not just where but how social scientists come to know about natural sciences, through the intimate relationship between laboratory studies, practice, and constructivism. The description of what really goes on in laboratories – the careful scrutiny and compelling revelation of the 'intricate labour,' the 'countless nonsolid ingredients,' the endless processes of 'confusion and negotiation' that constitute a 'fact' (Cetina 2001, 148) – has furnished science studies with its distinctive approach. (Garforth, 2012, 267)

Garforth's concern in 'In/visibilities of Research: Seeing and Knowing in STS' is on how close observation of actual practices reified the visual as the source of some kind of truth about how knowledge was being produced. Initially at least, this emphasis on making visible practices that had previously been little known was an important aspect of making visible and valued the 'invisible labour' often performed, work which is often inflected with gender, class and ethnicity power structures: 'The rhetorical emphasis is on seeing close up, in context, and in the middle of the action' (Garforth, 2012, 269).

Garforth draws our attention to recent scholarship that has tried to think outside the laboratory, or at least to rethink the limits of this. Studying data management is somehow outside the boundary lines of what counts as An Experiment, rendering it invisible in the Big Science literature. Furthermore, as Garforth highlights, laboratory studies have focused so intensely on the visual observations that studying data management instantly poses methodological problems in terms of what to observe. Coming to the process so early that at least in the first one or two rounds of interviews, there was literally no equipment to observe being used, led me to actively seeking out opportunities to visit other facilities. The sensation of actually laying eyes upon a beamline tunnel or a synchrotron felt as if it helped to give weight to the data I discussed with the Group Leaders, to somehow anchor this virtual information to something 'real'. In that, I fell into the trap of other STS observers wanting to watch something with my eyes:

> My analysis draws on recent contributions to STS that have emphasized the need to move out of the laboratory and explore new epistemic spaces that are characterized as diffuse, mobile, and hybrid. They call for the extension and reformulation of the canonical lab study methodology … to acknowledge the increasingly virtual and multisited nature of contemporary epistemic networks (Beaulieu 2004, 2010; Hess 2001; Hine 2007). (Garforth, 2012, 265)

The critique that Garforth develops of laboratory practices has been helpful in achieving a critical distance to black boxes in my own study. The encouragement to 'move out of the laboratory' in this article resonates with my interest in redefining the boundaries of what counts as The Experiment, arguing that data management practices play an important role in knowledge production that is currently neglected.

In the above I have summarised the early emergence of the black box, as well as the critiques of, and responses to, it. Thanks to these adaptations and discussions, it remains a useful tool for examining a part of a system which is typically removed from the view of users, such as the data management systems of the ESS. The black box itself is not a stable term, and the diversity of ways in which it has been adopted/adapted suggest that many scholars have found it a useful starting point for critically examining different instances of science and technology. In the context of this study, it has been particularly useful as a way to explore the in/visibility of the data management technologies.

Invisible work

In this book I am concerned with the co-constitutive relation between technological artefacts, human work practices and organisational infrastructures. While the black box draws our attention to development of artefacts that are typically non-human (for example, a fact, or a piece of technology), the concept of 'invisible work' highlights ongoing human practices, often in organisational contexts. Both are attuned to lifting up work that is often less visible. In this section, then, I turn to 'invisible work' as a way of examining the work performed by the Group Leaders of the Data Management and Software Centre (DMSC).

The concept of invisible work typically refers to work that is 'less visible' by virtue of being unrecognised or unvalued but which, at the same time, is necessary (Daniels, 1987; Ehrlich and Cash, 1999; Star and Strauss, 1999). It has been used to discuss a wide variety of practices, and as a way to start critical discussions about whose work and what kind of work is valued and why. One of the primary areas researched has been care work, including both that done formally by nurses, for example, but also the many daily, unremarked activities involved in caring for an elderly relative or small child (Daniels, 1987; Lydahl, 2017; Rio Poncela, 2021). As such these analyses often draw attention to the gendered and classed nature of such work.

Star and Strauss, writing within the field of computer-support cooperative work, are sensitive to the particular challenges involved with technically oriented work taking place in organisations that is often framed as support services, for example digital librarians or computer programmers. As Star and Strauss neatly explain in their seminal paper, 'Layers of Silence, Arenas of Voice: The ecology of Visible and Invisible Work', the concept of 'invisible work' brings a critical attention to: 'what exactly is work, and to whom it might (or should) be visible or invisible' (1999, 10). They also highlight a number of important characteristics of the concept, including flexibility (the concept can be applied to study many different kinds of in/visibility and acknowledges that levels of visibility shift over time and place) and the potential dangers inherent in making some kinds of work more visible (or that there are times when invisibility can be useful).

The work carried out at the DMSC is visible in a formal sense; the unit appears in organisational charts, it has a budget and targets of its own, its employees have job descriptions, annual reviews and take holiday time, its members regularly interface with other parts of the organisation and represent the organisation in certain public or professional settings. They have a webpage and their own email addresses. They organise their own conferences and take part in their own communities of practice. Therefore, the work carried out by the DMSC is not 'invisible' in the sense of being unpaid or informal or without status. Rather, I use the term here to explore how the work of data management is made 'invisible' in certain contexts and what might be at stake here through examining the intersection of data management and organisation. One possibility on the level of the scientific community is that the work of data management needs to be rendered invisible in order for the data produced to be considered as 'raw' and thus as the objective foundation for production of scientific results. However, the reasons why this work may be less visible may be somewhat different from an organisational perspective. Here, it might be considered as an example of what Star and Strauss describe as 'disembedding background work' (1999, 20–21) where the workers are visible, but the work they carry out is part of background expectations. In this way the work may be rendered invisible as a sign of the low value placed upon it. In these contexts, efforts to make the work more visible may be related to attempts at increased professionalisation by a particular group, or a desire to no longer be part of the infrastructure but rather to acknowledge a distinct contribution. More precisely, as the histories of Big Science mentioned earlier suggest, scientific computing and particularly data management work has previously been framed as support work to the experimental practices. Read through this lens the context of data management in Big Science fits well with the literature on infrastructuring.

Susan Leigh Star made a powerful argument for paying close attention to infrastructure in 'The Ethnography of Infrastructure', when she compared the wires and settings of an information system to the sewers of a city; these often-overlooked aspects are nevertheless vital for a well-functioning infrastructure.

> Study a city and neglect its sewers and power supplies (as many have), and you miss essential aspects of distributional justice and planning power (Latour and Hermant 1998). Study an information system and neglect its standards, wires and settings, and you miss equally essential aspects of aesthetics, justice and change. (Star, 1999, 379)

This is also true when studying so-called Big Science facilities such as CERN where, both historically and today, the 'invisible work' of data management systems form an essential part of the infrastructure necessary to producing cutting-edge scientific knowledge. Read this way, one might say that scientific computing (including data management) has been sidelined in scholarship about Big Science and is commonly understood as not being part of the experiment but rather a part of the infrastructure. This means that we lack understanding of the role played by data management in the knowledge production process.

Catching a glimpse of such infrastructures is often the greatest challenge as they become invisible when working well. This contributes to the sense that infrastructures are stable. The notion of 'infrastructures' has been revealed to be far from ideal in that it suggests such structures achieve some kind of stability over time, rather than remaining dynamic and responsive. As such, the recently coined 'infrastructuring' is a more useful term here, as it moves emphasis from the structure itself, to the practices performed in order to materialise the structure (Karasti and Baker, 2004; Karasti and Syrjänen, 2004; Star and Bowker, 2006; Pipek and Wulf, 2009; Karasti and Blomberg, 2018). In the following section then, I turn to how this has appeared particularly in relation to movement and management of information.

Informational infrastructuring

> scholarship in sociology of science and science, technology & society has shown how infrastructures organize the circulation of knowledge in society (Edwards, 2010; Borgman, 2007; Bowker et al, 2010). (Plantin et al, 2018)

Significant amounts of scholarly expertise have already been devoted to informational infrastructures, sometimes referred to as 'cyberinfrastructures'. This body of knowledge has explored how in different empirical contexts production of scientific knowledge is dependent upon and shaped by the digital infrastructures that move data around between different (usually human) entities (Paine and Lee, 2020; Karasti, Baker and Millerand, 2010). Studies have explored how infrastructures for large e-Science projects involve both researchers and developers. Work has also examined the different kinds of more or less explicitly political demands placed by the institutional context (Nost, 2022).

A recurring theme in the CDS literature previously discussed is the 'invisible' nature of the work done by digital infrastructures to collect and process data, whether this manifests in a belief that data holds no traces of these processes, or a lack of value ascribed by an organisation to this kind of work, which is often perceived as part of the infrastructure. My hypothesis is that data management systems in Big Science are an example of infrastructuring, which is often treated as 'invisible, part of the background for other kinds of work' (Star, 1999). It can therefore be challenging to make this visible and explore how this kind of infrastructure might shape knowledge production.

As Seidel's historical account of the development of computing resources at FermiLab shows, demands for computing resources capable of managing increasing amounts of data were entangled with budgets, departmental politics and the paradigm shift from supercomputers to computer farms. Data management has thus long been a core part of the organisational infrastructure at Big Science facilities. Studying big data and data management tools can shed light on the organisational infrastructure of Big Science facilities. Capture, processing, storage and movement of data has practical implications in terms of hardware, software, staff, and premises. Furthermore, big data may increase the visibility of particular teams by placing greater emphasis or demands upon the technical skills associated with data management (Buckland, 1989). Big data poses new challenges for software in terms of volume and complexity, and consequently drives the development of information and communication technologies (ICT). This also has implications in terms of the creation of physical space in a facility for computing centres as well as a requirement for training programs required to teach scientists to use the data management software at that facility (Hanseth and Monteiro, 1997). Decisions about how to structure an organisation reflect the specific circumstances under which they were made. These surrounding 'social processes' are thus embedded and embodied in the information systems that make up the core of a modern organisation's infrastructure: 'The technical basis for an information infrastructure is the standards which regulate the communicative patterns. *These standards are currently negotiated, developed and shaped through complex social processes.* They embody inter-organisational changes in the specific way they regulate the communicative patterns' (Hanseth and Monteiro, 1997, 183, my emphasis).

Bowker and Star's work (2000) on standardisation is a prime example of how to study procedures necessary for the smooth functioning of an institution (in their case the International Classification of Disease (ICD) created by the World Health Organization) in order to explore assumptions coded into processes for capturing, processing or sharing data. Their study draws attention to both the broader historical and socio-cultural contexts, and also the level of technological development, suggesting that analysis of informational systems should never be conducted in a vacuum but rather with awareness of the contingency of the format and content of such systems.

Data management systems, and surrounding informational infrastructure, comprise an integral part of the functioning of New Big Science (NBS) facilities, one which may be less visible than the experiments themselves but which nevertheless can tell us a great deal about the processes and priorities at work within the facility. In particular, successful use of these systems is a badge of membership of a particular community of practice (Wenger, 1998), one that excludes non-specialists. Data management software and hardware are an essential part of the infrastructure of a new Big Science facility, albeit one that is rarely considered until it breaks, but competent use of which signals membership of a particular community of practice. Studying data management in these facilities can thus produce crucial insights into how practices at 'ground level' may be inflected by broader politics as well as the changing understanding of data management within NBS. How then to make visible these practices which often disappear into the background?

The infrastructure organised around/supporting data management systems assumes/proscribes a certain way of working with data (Akrich, 1992). However, the interplay between infrastructure and information management often occurs in the background making it hard to study. Following the paper trail of development documents can be dry work, and the kind of everyday 'hacks' that make a system work correctly are often poorly understood or undocumented. Despite these difficulties, adopting a mixed methods approach that incorporates system analysis with analysis of policy documents, interviews and observations can provide an important entry point to understanding how decisions about data management reflect institutional priorities or shape the ways in which data is captured. Star (1999) suggests the technique of 'surfacing invisible work'. This approach brings attention to who is doing work that contributes to the smooth running of the facility, but may not be considered to be 'doing science'. In these terms, many aspects of data management might still correspond to 'invisible work' and would repay increased attention from researchers.

We can thus learn a great deal about how a facility functions by examining its informational infrastructure. Various entry points to exploring the big data infrastructure include asking: how is information traffic prioritised? What is archived and for how long? Who is responsible for doing this? How much technical support is available for different computing functions? Is the code open source or proprietary? How visible are the technical staff responsible for computing in the facility itself and what are their backgrounds? How do users learn about the software? What is the process for making updates to or fixing software? How does the system interface with other (external) systems?

How do scientific computer experts work together with instrument scientists to capture data from experiments? Examining aspects such as these can reveal, for example, internal power structures and relations between different parts of the organisation, as well as openness to external organisations.

As Star acknowledged, studying organisational infrastructure is not particularly sexy or visible. It often involves examining rather dry documents or routines that appear to be secondary to the 'real' work taking place. Much of this less visible work (what Star aptly calls 'the forgotten, the background, the frozen in place' (1999: 379)) is now accomplished by or in tandem with information technologies. Buried in the digital protocols and routines there are clues as to how power dynamics and knowledge production are organised at 'ground level', how the organisation of office space, the development of standardisation or categorisation protocols, or access to technical support reflect and shape inter-organisational power dynamics. A technique known as 'infrastructural inversion' has been used as a way to disturb assumptions about changes to organisations through focusing on structural changes, as Blok illustrates:

> For instance, whereas it is widely assumed that advances in medical science caused the rise in life expectancy during the nineteenth century, performing an infrastructural inversion will show that changes in living conditions, tied in particular to improved diet and sewage, were at least as important (Bowker, 1995, p. 235). (Blok et al, 2016, 10)

This quotation appears in the introduction to a special issue of *Science As Culture* about infrastructures. In it, the authors discuss the technique of 'infrastructural inversion' in which one starts by examining those aspects which are easily overlooked or appear only as part of the background. In the previous quotation Blok et al give the example of life expectancy. In this case, the dominant view of changes to life expectancy is challenged; rather than medical science being the 'hero' of the piece, the more mundane incremental improvements to diet and sewage are seen as being equally important. This infrastructural inversion is a useful example in that it gives weight and visibility to the less glamorous practices. If we were to transpose this example to the world of Big Science, a world characterised by big promises of world-changing advances in scientific knowledge driven by the brightest source in the world, then I begin to wonder if the quiet work of data management might be just as important to world-changing knowledge as the experiments carried out by visiting scientists? If I am to perform the same trick on Big Science, we might focus on how advances in scientific knowledge production might be tied as much to developments in data management as to more powerful sources that allow us to 'see' more. This upside-down view of the world of Big Science is a useful trick in making visible the work of the Group Leaders at the DMSC, and to ask what are the consequences of the situated decisions they made when developing the data-management pipeline for DMSC.

The goal therefore, of this book is to examine the following proposition: the more-or-less 'invisible' practices of data management infrastructuring can – if studied carefully enough – tell us something about the ESS and the changing nature of Big Science. 'Invisible work' here refers not only to the absence of accounts of data management work in the literature about Big Science but also enquires into the in/visibility of those doing the work of data management infrastructuring: 'What needs our ethnographic attention, Star argues, are those relational settings in which otherwise invisible infrastructures become visible, not infrastructures per se, but the practices, materials, and settings of infrastructuring' (Blok et al, 2016, p 3).

With this in mind, I have paid particular attention to that which highlights the connection between infrastructuring and epistemic politics or the production of particular kinds of knowledge (Cetina, 1999; Doing, 2009). In doing so, I wanted to lay the foundation for Chapters 5–7 in which I (i) dig deeper into the technical infrastructure through the lens of alignment work; (ii) consider the changing role of technical experts such as my participants as creators of scientific knowledge, and; (iii) turn a critical eye to the infrastructuring that takes place when technical knowledge provided through the In-Kind Contribution (IKC) structure meets the political challenges of a multinational organisation like the ESS.

In summary, invisible work and the black box are used here as a way to surface infrastructuring as it takes place at the DMSC during construction. In so doing, I seek a deeper understanding of the epistemological and methodological assumptions behind experimental practices at the ESS through studying the data management infrastructures. One important question to ask, however, before setting off on such a journey concerns why such infrastructures have historically been less visible and what is at stake in making them more visible now?

What is at stake in visibility?

> we need to reflect carefully on the kinds of secrecy that surround specific knowledges and experiences of working practice and the implications of making them visible. (Suchman, 1995, 56)

While the black box and invisible work are undoubtedly powerful tools when exploring practices of infrastructuring in Big Science, an important question that also arises in the scholarship around these terms is: what is at stake? Who benefits from the 'back stage' becoming visible to the 'front stage'? (Suchman, 1995). What difference does it make that the DMSC's work is 'invisible' in the context of experimentation and producing scientific knowledge? Who benefits from that? What would be different if the DMSC work was visible and valued in the experiments? In trying to think this through, I returned to Star and Strauss' discussion of front and back stage:

> There is a special instance of embedded background work, which paradoxically may result in a highly visible public performance. Here Goffman's analysis of 'front stage' and 'back stage' is particularly compelling (1969). Many performers – athletes, musicians, actors, and arguably, scientists – keep the arduous process of preparation for public display well behind the scenes. Thus the process of trial-and-error in science is less visible than the final published results (Shapin, 1989; Star, 1989). (Star and Strauss, 1999, 21)

If we open the black box of data management by examining development histories and lift up the work being done at the DMSC by making visible the disciplinary backgrounds of my participants, what impact will this have on perceptions of the data being produced, and for whom? While greater recognition of expertise sounds like a positive outcome, could shedding light on how data is produced muddy the waters of scientific objectivity? Garforth, developing Star and Strauss' argument in relation to the trope of visibility makes a connection to accountability. In the context of contemporary knowledge production, where funding and bibliometrics are key, Garforth draws attention to the possible benefits of some work remaining invisible:

> Backstage can be a space for making mistakes, for processes of trial and error which are crucial to the development of competence but need not be submitted to the public gaze. As such, invisibility is closely linked to autonomy and discretion in work processes. Making practices amenable to scrutiny can make them count (acknowledged and valued) but also draws them into logics of accountability, either in relation to a specific observer or a generalized observing – and auditing – eye (Star and Strauss 1999, 9–10). (Garforth, 2012, 276)

The literature on invisible work also encourages us to not only pay attention to which processes are invisible, but to whom and why, when and where. This connects productively to different agendas around the production of 'raw' data. For whom is it important that the work around collecting and managing data disappears in order for the appearance of 'rawness' to hold for the data? For whom might this work be important to highlight as part of organisational structures and negotiations over resources (see also Scroggins and Pasquetto, 2020).

Both the black box and my other accompanying concept, invisible work, thrive on the imperfect, unfinished nature of the DMSC. Both concepts see value in the under-construction status of the site because it allows a glimpse into the decision making process, literally making visible the assumptions and tensions involved in developing a centre such as this. It is a period in the history of the facility when there is acceptance of messiness and incompleteness. Although the goal is of course to deliver a smooth-running data management system, at the time of my visits, I heard stories of things going wrong, of grumbles about the organisation, of surprises and changes to budgets and plans. These are the parts of the development which may be forgotten when the facility is up and running, and steps are harder to retrace.

Conclusions

In the introduction I proffered the notion of the black box as one way to illustrate how little is known/made visible about data management at Big Science facilities. As I will go on to show, data management comprises a whole series of black boxes offering an apparent treasure trove of artefacts to be 'unpacked' in order to understand what happens to the data, and to make visible the expertise and contribution of those who work in data management.

The most wonderful thing about this project was the chance to follow the construction of the ESS, to be there as the walls literally rose, as my participants moved from one set of premises to another, as the teams grew, changed and the leadership shifted from person to person. Unlike some of my previous research, this time I was able to observe as plans for data management were made and variously implemented. I had the chance to ask my participants about the impact of the rebudgeting exercises that happened during the three years of interviews. Together the black box and invisible work allowed me to follow the construction of the data management system as the combined result of a specific set of technologies, people and organisation. These concepts also prompted questions about what is at stake in making visible this context.

The conversations I had with my participants capture the dynamic, constantly changing situation. This makes the interview material very lively, but perhaps introduces some challenges for me in creating a coherent story about what happened. In the following chapter, I turn my attention to the materials and methods involved in carrying out this study.

4

How to Study a Hole in the Ground

The story of this research journey starts in 2014 with a funding application made to the Marcus and Amalia Wallenberg Foundation by a group of scholars at Lund University. Thomas Kaiserfeld, Mats Benner and Kerstin Sandell had, in various ways, already been engaged in exploring the impact of Big Science on Lund when they led the funding application that included money for a postdoctoral research project that became this study on data. The project reflected their diverse expertise but focused in on one question: *How do the structures and cultures of the New Big Science affect how research is practiced, funded and organised?* They argued, in the application, that there was something distinctive about this new generation of Big Science facilities, and, in the research outlined in the proposal, they proposed to investigate where this distinctiveness lay. When asked to join the team and sketch an interesting entry point, I chose to focus on data. In this chapter, I will discuss *how* I went about studying data at the European Spallation Source (ESS), connecting this discussion to the existing scholarship and laying out my methodological choices and the challenges I encountered along the way.

At the time of the application, the idea of 'big data' had already become common parlance and discussions about 'big data' were highly visible in both scholarly and popular publications. Mesmerised by the vast amounts of data that facilities like the ESS produce during experiments, the high stakes of the game being played, and against such a backdrop of 'buzz' about data, I was puzzled as to why there was not so much attention to data management in Big Science. I had read the histories of other facilities (see Chapter 2 for a discussion of these) and management of experimental data was curiously invisible in many of these. Various volumes were being produced about the ESS already, perhaps unsurprisingly given the long lead time before construction started, and data was covered only briefly in these. From what little I had found about data management in Big Science as it had been carried out historically, this work appeared to be poorly funded and supported at an institutional level, and instead was often carried out within individual research teams in terms of analysis and processing. The ESS looked different for a number of reasons: dedicated site and staff for management of experimental data, clearly allocated funding and providing support for the entire data pipeline (from experiment to analysis).

With this in mind, I outlined my part of the project and – when the funding was granted – I then started to consider how I might study this. This chapter provides a detailed account of the fieldwork that was undertaken, including materials, methods and reflections on the role of the researcher. It includes samples of the interview questions that were used during conversations with the participants to organise the dialogue around three themes: technologies, people, organisation. Finally, it discusses the challenges that arise when studying something 'when the box is still open'.

The Group Leaders

Here I explain the structure of the Data Management and Software Centre (DMSC) when I started to follow their work, and how that shaped the materials I chose/was able to collect. The ESS has two locations, one in Lund, Sweden, and one in Copenhagen, Denmark. The DMSC is located in Denmark. In 2015 they were located in temporary premises in an old building of the University of Copenhagen, although they were looking for a new, permanent home. Meanwhile, the ESS site in Sweden was in the early stages of construction and mostly resembled a building site.

As of spring 2023, the DMSC's remit is presented as follows on the ESS website:

> DMSC designs, develops and supports the ESS scientific data pipeline, including experiment control, data acquisition, data curation, scientific web applications, data reduction, data analysis and modelling, data systems and data centre operation.
>
> Our objective is to enable and support a high impact science programme across the ESS neutron instrument suite. (ESS, 2023; Data Management & Software)

At the DMSC the work is organised into different groups, each of whom is responsible for a different aspect of the data management process detailed earlier. Group Leaders have been recruited from other Big Science facilities and major software companies. They lead small teams (typically between three and seven people) of ESS employees as well as coordinating a wide range of external contacts and contractors[1] through the 'In-Kind Contribution'

[1] The primary source of external expertise is through the In-Kind Contribution model through which other European countries are able to contribute equipment and

(IKC) model, European Union projects and other collaborations. At the time of my first contact with the Group Leaders, the groups were called: Instrument Data Group, Data Systems and Technologies Group, Data Analysis and Modelling Group, and Data Management Group. As the names suggest, each group had a different area of responsibility, for example the 'Data Systems and Technologies Group' was responsible for developing the hardware, including provisioning a server room. While the 'Data Management Group' were responsible for the flow of data from source to scientist. There was also a DMSC Project Coordinator, who oversaw planning for the DMSC. Together with the Head of the DMSC, these people were the core group leading the development of the DMSC itself and the services it would provide for the ESS.

When I started following the DMSC in the Fall of 2015, the unit was still quite small and their work was in the early planning stages. Despite this the Group Leaders already knew that there would be a significant increase in staff, a change in premises and materialisation of specific pieces of software and hardware in the next few years. With so much anticipated change in the time span of my project, I opted to focus on the same core group of people to interview. Focusing on people at management level within the DMSC also allowed me to learn much about the surrounding context thanks to their prior experiences. They were ideally positioned to reflect on organisational dynamics, as well as being skilled at explaining the technical terms to non-experts such as myself. Returning to talk to them every year gave some important consistency to the fieldwork in the midst of so much change (or 'movement' as I called it in the Introduction). These conversations formed the core of my study and in the chapters that follow I have endeavoured to give as much space as possible to the Group Leaders' voices. This study would not have been possible without them, and their distinctive experience and knowledge has inevitably shaped my understanding. For this reason, I also refer to them as 'participants' rather than 'interviewees'.

The Group Leaders were responsible for not only managing a group but also for developing the vision or solution for ESS for their area of responsibility. Their remit was both internal-facing to their team (including a busy recruitment program) and the rest of the ESS organisation but also external-facing as they nurtured the IKC contacts user groups and other stakeholders. My instinct was that these people would be the best ones to follow in order to learn about the development of data management tools and techniques at the ESS. Or at least they would be a good starting point.

---

expertise to the ESS project. This model is discussed in greater detail in Chapters 5 and 7.

I approached the Group Leaders in early Fall 2015 by writing individual emails in which I introduced myself and my project. I also requested a 45-minute interview with them. In the first year five of them agreed, comprising Leaders of each of the four groups, and the DMSC Project Coordinator, but not the Head of the DMSC at that time. I supplied them with an information sheet about my project and a consent form. In this form, I requested permission to record the interviews and publish my materials. I also asked them to waive their anonymity in this research on the grounds that their disciplinary backgrounds and professional experience were an important part of understanding how they approached data management. Given the small group and the relative size of the community in which they work, I felt that it would be impossible to preserve their anonymity in this project. Fortunately all agreed.

Let me introduce you to them:

Mark Hagen	Head of DMSC Division (until end 2016)
Petra Aulin	DMSC Project Coordinator
Thomas Holm Rod	Group Leader for Data Analysis and Modelling
Jonathan Taylor	Group Leader for Instrument Data (Acting Head after Mark Hagen, then Head of DMSC)
Tobias Richter	Group Leader for Data Management
Sune Rastad Bahn	Group Leader for Data Systems
Afonso Mukai	Data Management Group

In Chapter 6 I discuss the backgrounds of the Group Leaders and their different trajectories to the DMSC.

Interviews

In the Fall of each year (2015–17) I carried out semi-structured interviews individually with each Group Leader. My aim was to interview the same people, asking about the same topics to track development of data management as the facility was being constructed. The interviews typically lasted 1–1.5 hours. In the last two years (2016 and 2017), the interviews also involved eliciting diagrams from the Group Leaders, which were primarily used as an aid to help me understand the highly technical work that they do. The interviews took place at either the DMSC buildings in Copenhagen or at the temporary ESS building in Lund.

The interviews were semi-structured, recorded using a digital voice recorder and took place in meeting rooms in DMSC/ESS premises. Given the differences in their roles, the questions I posed to the Group Leaders were designed to be fairly open, with scope to follow interesting lines of conversation as they arose. In the first year I asked all of my participants the same opening question: 'What's your background?' In the following years I opened all of the interviews by asking what had changed since the previous year. Throughout there was a common theme of looking forward and backwards. For the individual Group Leaders I often spent some time before we met reading the latest press releases on the ESS website to see what developments had taken place that might impact their work. We often discussed how work at the ESS compared to other Big Science facilities, or to their experience in commercial software development.

These interviews were recorded, transcribed and analysed using a tripartite scheme of artifacts/practices/arrangements (Lievrouw, 2014), allowing me to investigate data management as a situated practice of knowledge production and mediation. The interviews comprised a series of questions organised around three themes:

- Arrangements – disciplinary paradigms and other organisational patterns of working/analysing/understanding (context or production).
- Practices – who are users and what do the Group Leaders imagine they are going to do with data, also where do these ideas about users and their needs come from (uses and audience)?
- Artifacts – database and data itself containing affordances and limitations (technologies or texts).

These more abstract-sounding themes became the foundation for specific interview questions that focused our conversations. I would often open the conversation by inviting participants to:

- Look forward and backwards – placing development of the data management systems within a chronological framework focused on development of the ESS organisation.

Often their narratives about the development of the ESS prompted questions and reflections about their own role:

- Who are you, what do you know and how do you expect these technologies to be used – examining professional expertise, experience, transfer of knowledge between people and facilities, and accompanying user assumptions.

Specific pieces of equipment emerged as illustrations in the discussions mentioned earlier and presented an opportunity for me to learn more on a technical level:

- Tell me about the technologies – learning about limitations and affordances of hardware and software.

Close reading and rereading of the interview transcripts allowed me to identify passages which responded most closely to the three themes. Once identified and organised by year and participant, I zoomed in on them to analyse *how* the participants talked about their work at the DMSC. These three themes produced the three central chapters in this book as follows:

- 'Getting Technical' (Chapter 5) is about the technologies themselves.
- 'Technician or Scientist?' (Chapter 6) is about the people.
- 'Organisational Frictions' (Chapter 7) is about organisational (infra) structuring both inside and outside the ESS.

While there is empirical material used throughout this book, Chapters 5–8 give the most space to the participants and their reflections around these themes. For the sake of clarity these themes are separated into three chapters (5–7), but it is important to stress that they are also very much entangled together.

Additional materials

The interviews I conducted with my participants form the material at the heart of this book. The ESS was in an early stage of construction during the fieldwork period, which meant that there was very little to 'see' in the sense of a traditional ethnography. My participants worked in office space in first a temporary location, and then in a building shared with other companies. Consequently, our annual appointment for an interview became the richest resource available to me. However, I also collected a number of documents, including press releases and activity reports from the ESS website, and other key documents freely available online, including the *Statutes of the European Spallation Source ERIC* and *ESS Technical Design Report*. These documents, particularly the activity reports and press releases, gave a glimpse into how the ESS was being represented to the public. I often used these as a way to prepare questions or anticipate conversation topics for my interviews by checking on what had happened at the facility since my last visit. Later in my study, and as I wrote this book, I also collected academic journal articles, presentations and project reports co-authored by some of the Group Leaders which gave more detail on the technical visions and outcomes prior to and after my period of fieldwork. These included, for example, *EU SINE2020 WP 10: Report on Guidelines and Standards for Data Treatment Software* (a report from an EU project designed 'with the objectives of preparing Europe for the unique opportunities at the European Spallation Source'), as well as a

2015 article titled 'Hardware Aspects, Modularity and Integration of an Event Mode Data Acquisition and Instrument Control for the European Spallation Source (ESS)' (Gahl et al, 2015). These documents were primarily used to support my understanding of the technical setup, and illustrations from such reports appear here as aids to the reader to explain the complex process of data management. I also drew diagrams during interviews, often with assistance from my participants. These were primarily used to help me understand the technical setup, but I also include one such diagram in this book as part of my discussion on aligning of different technologies in Chapter 5. Last but not least, during the period of fieldwork the DMSC moved from one building to another in Copenhagen, and I took photographs of both premises. These appear in Chapter 7. Following the end of the formal fieldwork period, I returned to the DMSC on two occasions in 2018 and 2019 to discuss a draft text and present initial results. These were important opportunities to clarify and correct misunderstandings, as well as hearing the reflections of the Group Leaders on my findings. This helped to frame the discussion in Chapter 8 in particular.

In February 2017 I visited ISIS and Diamond Light Source, both UK Big Science facilities, and in December 2017 I visited MAX IV, a neighbouring facility to the ESS. Through these visits I sought to gain a broader perspective on data management in Big Science by learning more about other facilities, which were well-established and often which had links to the ESS DMSC team. During these visits I conducted semi-structed interviews with members of staff with different kinds of expertise, including Software Engineer for Data Acquisition Group, Principal Scientist, Group Leader for Research Data, and Service Manager and Developer for Scientific Computing. In addition to our conversations, they gave me tours of the facilities and I had the chance to present and discuss my own work with them. These visits comprised one or two days at each facility, rather than an in-depth ethnographic study and were intended to supplement my primary focus on the ESS by providing a glimpse into how data management actually functioned at working Big Science facilities. Timed as they were before my final round of interviews at the ESS, they also informed the questions I posed to the DMSC Group Leaders thanks to the bigger picture they had provided. For example, I was able to appreciate more clearly what was novel and different about the ESS, and what was a continuation of existing practices in that professional community.

Limitations of the study

Although I started with an open mind about what data I would collect, I quite quickly resolved to focus on the Group Leaders and their experiences. This was both a pragmatic choice to achieve consistency across the course of the project but also a choice driven by personal interest. During the first round of interviews we talked longer than I anticipated and the conversations around data were very fruitful. I also felt quite strongly that these were voices that had not been heard in other accounts of Big Science facilities and the notion of 'invisible work', understood as the necessary work that must be done to keep something running but which often goes unnoticed (see Chapter 4 for a detailed discussion of this), emerged early on from these conversations as a useful lens for studying data management in Big Science.

I chose to stick close to the Group Leaders thus for reasons of consistency but also because I found it fascinating to talk to them and see another side of Big Science which hadn't appeared in the other accounts I had read of this field. I suspected that their personal career trajectories were shaping how they approached the technical challenges of building up the DMSC, as well as their reflections on the emerging organisation. As such, I looked through their eyes at the organisational politics of ESS, and I am sure I would have had a very different perspective on data management had I chosen to interview other groups at the ESS.

Another important limitation is of course the time period. Construction of the ESS is still ongoing as I write this, and the COVID-19 pandemic caused delays such that the final completion date has been revised. My study concerns a very early period in the DMSC development when decisions were being made, but no data was actually being captured from experiments. The difference between the construction and operation phases of such a project is likely to be significant but is outside the scope of this book unfortunately. It is also important to note that even when the facility moves into full operations (and the black box of data management appears closed), the work of the DMSC will by no means be complete. Indeed, it was partly because of this that the notion of 'infrastructuring' (Karasti and Blomberg, 2018) in relation to the 'invisible work' of data management became so relevant.

Finally, it is worth noting that even following the ESS during construction did not guarantee access to the full history of all the technologies I encountered. There are layers and layers and layers of hardware and software sifting and processing the data. And every one of these has a different provenance, a different set of baggage. Back in Chapter 2 I introduced Latour's notion of the black box as a way to study technologies. I was particularly interested in his note that 'It's easier to be there before the box gets all black' and to see following the ESS during construction as thus a golden opportunity to be there 'before the box gets all black'. I felt optimistic that I would be able to explore the thinking that went on around the design and development of the data management system for the facility. What I discovered in reality – and which is reflected in Figure 5.2 in Chapter 5 'Getting Technical' – is that while the system itself was open to exploration, there were still a number of very black boxes that fulfilled key roles in the system. For example, Kafka or

Mantid, a piece of software that (although not ultimately selected for use) was touted as a solution for part of the DMSC system for several years. Both of these predate DMSC and have been significantly developed outside ESS for use in other application areas. They thus contain other development stories to which I did not have access, and are the results of long term collaboration.

Challenges: how to study that which is under construction?

The interviews were essential in lifting up the work of 'infrastructuring' as it took place, and before it disappeared from view as organisational structures stabilised, or at least appeared to be less obviously contested. By returning every year to talk to the Group Leaders I was able to ask questions about what had changed since our previous meeting and what they anticipated in the year ahead. This had two advantages: (i) gaps between plans and realities became clear (Suchman, 2007); and (ii) the many small steps or negotiations that take place (and which are often forgotten when the 'finished product' is in place) are revealed. Decisions that feel significant at the time may become obscured by the passing of time and the volume of other decisions and events surrounding them. By virtue of meeting regularly with the Group Leaders, some of these smaller steps were recalled and discussed. However, as I hinted at in the end of the previous section, this under-construction phase also made it harder to step back, and get a clear sense of what the biggest debates and questions for data management were. When everything is important, it is hard to see what will be the lasting impact. Or, to easily define a research 'object'. In some ways, writing this book some years after fieldwork has assisted me in that task. Due to various periods of parental leave, not to mention the pandemic, there was a delay between fieldwork and the sabbatical that allowed me the time to finalise my writing. In this gap of time, I found it easier to step back and get some perspective on my interviews. It became easier to trace themes across the years, to see possible patterns.

A more practical challenge associated with studying something that is under construction was what to actually look *at*. Unlike other Science and Technology Studies (STS) studies of laboratories (see Chapter 3 for an overview of this practice in which some kind of informal 'hanging out' at the lab waiting for something interesting to happen is quite common) much of this study took place before there was a physical lab, or a server room or even a proper, permanent home for the DMSC. By necessity my study had to be flexible enough in design to move and follow the DMSC as it literally moved. While hanging out in a lab allows for the grainy realities of everyday decisions and negotiations to be captured, interviewing less often but in a more reflective vein allowed me to capture important details without getting hopelessly lost in the planning that was taking place. This proved to be particularly helpful given the technical and organisational complexity of the ESS project.

Conclusions

In focusing so closely on the Group Leaders I aimed to provide an account of the 'invisible work' of data management in Big Science, a topic neglected in the social sciences-based accounts of Big Science but which is fundamental to the production of scientific knowledge. Through regular interviews and collection of supporting materials as the ESS was being constructed I aim to give a glimpse into one aspect of the making of the facility, before the particular 'box' of early-stage data management is closed. My methods were qualitative and provided a detailed account grounded in the experiences of a small group of people. My approach is broadly in line with much research conducted within STS. It is perhaps, however, also important to note that – as a gender studies scholar by training – my fieldwork and writing process has a distinctive flavour.

This book is about exploring one particular part of the context surrounding knowledge production as it takes place in Big Science, more precisely the management of experimental data that will take place at the ESS. In lifting up the specificity of the context, I want to ask what effect (if any) it has on the production of 'raw' data. This is the story that concerns most of this book. However, in order to answer my research question I rely on the 'raw' data that I collected, which took place under specific epistemological and methodological conditions. These are what might be called the 'context' of my own knowledge production.

Given my own ambivalence about the promise of 'raw' data as the foundation for scientific knowledge production, and the growing popularity of CDS approaches, it would be too easy to take on a straightforwardly critical position towards the world of Big Science. By making clear the 'situatedness' of my own study (Haraway, 1991), I want to be clear about the limitations of the research I did – not to invalidate it, but rather to make space for alternative perspectives. To say, this is the view from here. Other views may be available. And most particularly, to make space for my participants' perspective.

I knew, from the very start of the project, that I would likely have a very different disciplinary background to the people I hoped to interview. I knew that I might be an unlikely intruder into their world. I also knew based on previous experience, that the epistemological foundation in which one has been raised tends to produce a view of the world which feels 'natural' or 'obvious' to oneself, and that encountering other worldviews maybe unsettling.

In order to address these challenges, I approached the field as an exercise in learning (Nielsson and Svensson, 2006) and many of the conversations involved the Group Leaders explaining to me how the various technologies worked. I was lucky that they were skilled and patient pedagogues as my lack of technical training meant lots of questions. This also gave a particular flavour to the study; these were conversations more than interviews, participants not interviewees, which made space for an acknowledgement of the coexistence of very different ideas about data. This emerged in the interviews, but also other kinds of conversations. I wished to include the voices of my participants as much as possible, and give them the chance to comment on my work while I was drafting and analysing. This took the form of meetings with the Group Leaders to share texts or report my early findings, with the invitation to discuss. It is also the reason why I include sometimes quite lengthy excerpts from our interviews in Chapters 5–7. This part of Big Science has previously been invisible to many, and including the voices of my participants is one way in which I hope to address that imbalance and recognise the work that they do.

5

Getting Technical

The role of the Data Management and Software Centre (DMSC) covers a broad remit, from initial idea through to publication: 'a scientific computing division that provides the services and solutions that users need for performing an experiment. This encompasses a fully integrated data pipeline from proposal to publication' (DMSC, 11 April 2023). This is what we might call the 'promise' of the DMSC (Dimitrievski, 2019). The broad scope reflects the changing expertise of users, as well as an ever increasing push to publish within academia. What does this pipeline look like? And how much of it is visible to the users? The following extract from the technical design report published in 2013 provides a helpful overview:

> To fully exploit the information power of ESS, a new approach for software and data management is needed that intuitively integrates control of the neutron instrument and its sample environment; data processing, visualisation, analysis and publication; and permanent storage and public access. Realising this vision of a fully integrated e-science solution from idea to publication will be one of ESS's major contributions. The ESS-Data Management and Software Centre (DMSC) will tackle this e-science challenge, delivering a 24/7 e-science service programme to cover the complete research cycle from idea to publication. (Åberg et al, 2013, 132)

As Figure 5.1 shows, *data management* is just one aspect of the work that will be performed by the DMSC when the facility is up and running. Historically users visiting facilities like the European Spallation Source (ESS) had the beamline scientist as their primary point of contact. These people were the local experts on the beam, familiar with the quirks of the line, responsible for providing training for new users and answering queries. They also provided invaluable support in terms of making sense of the data that was generated, or providing models with which visiting scientists could

Figure 5.1: Data research cycle

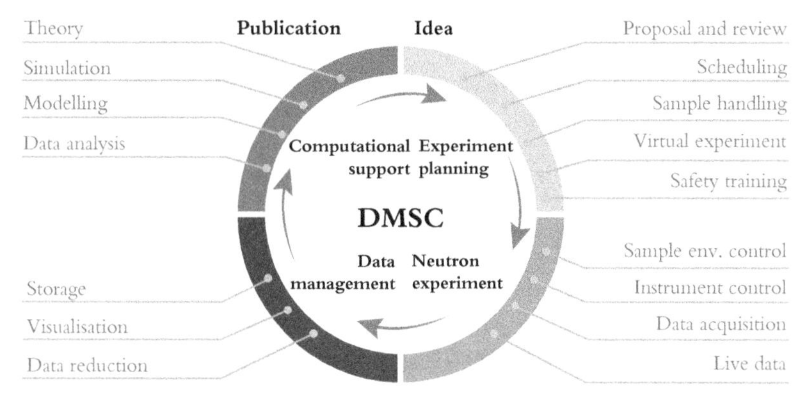

Note: This diagram was included in the technical design report for the ESS and shows the various different stages in the data lifecycle where the support and expertise of the DMSC would be required.

Source: Åberg et al (2013)

compare their own results. This significant contribution to the experiment meant that many beamline scientists appear as co-authors on publications. The DMSC, however, work on a different layer, one step removed from the individual experiments. How then can we understand their role in the production of scientific knowledge?

The Group Leaders I met are responsible for different parts of the life cycle, but must work closely together to ensure that the data flows smoothly from experiment to end user. Meeting with Jon Taylor in 2017, this focus on connection was very clear:

Jon:	The technology is, those decisions I think more or less are made. The kind of groundwork, the foundation of all the bits of, you know software, hardware. Those decisions are made. The challenge really is to turn the thing that's the design on paper into an operating facility. Really that's the challenge.
Katherine:	So this is making it materialise.
Jon:	Well I mean, yeh making it work. Making all the bits work together. (Interview with Jon, 2017)

The challenge of 'making all the bits work together' was a common thread across my interviews. How to carry out this alignment work and the impact it had on the data form the focus of the remainder of this chapter, providing a window into the technical details of the work done at DMSC while at the same time maintaining the focus on the journey taken by the data.

In this chapter I use Critical Data Studies (CDS) perspectives to look in depth at the trajectory of data collection; from experimental sample to live data stream to archiving/analysis/visualisation involves many different pieces of hardware and software. How do the DMSC staff align the different technologies to ensure data validity at the ESS? To answer this question, I want to dig deeper into the specifics of the technical infrastructure being constructed at the ESS to handle data collection and management. I will introduce some of the different kinds of software and hardware discussed by my participants and the process of aligning these heterogeneous components to ensure safe delivery of data to visiting scientists. Drawing on CDS, I will use the concepts of the 'data journey' (Leonelli) and 'data friction' (Edwards) to situate the movement of data as a series of sociotechnical decisions about what data is 'valid' and to make visible the work done together by people and machines to collect, safeguard and deliver 'raw' data. These lenses allow me to suggest some different ways in which data may be situated through localised human-technology collaborations. But first, it's time to say a little more about one of these technologies, which so far in this book have been at a distance.

Introducing Kafka

During 2016, the arrival of a new, central piece of equipment called 'Kafka'[1] was a hot topic of conversation when I met with the Group Leaders. Given both the enthusiasm that it generated among participants and its role in the kind of alignment work mentioned earlier, it quickly presented itself as an ideal focal point for discussing the challenge of 'making all the bits work together':

> Apache Kafka is an open-source distributed event streaming platform used by thousands of companies for high-performance data pipelines, streaming analytics, data integration, and mission-critical applications. (Apache Kafka webpage)

> Kafka is a solution to the real-time problems of any software solution, that is, to deal with real-time volumes of information and route it to multiple consumers quickly. *Kafka provides seamless integration between information of producers and consumers* without blocking the producers

[1] Kafka is a commercial solution used by many companies that need to handle large volumes of data. Perhaps its best known customer is LinkedIn.

> of the information, and without letting producers know who the final consumers are. (Garg, 2013, my emphasis)

In brief, Kafka collects data in one place and makes it possible for those who are interested in the data to read it, by subscribing to its messaging service. As part of this it provides temporary storage of data and replication of data. Kafka is particularly good at securely handling large volumes of data in real time. It is also a scaleable solution so appropriate for organisations of different sizes. Kafka is a key part of the data management at ESS. In my attempts to understand how Kafka works in relation to the rest of the equipment, I sketched the following diagram (see Figure 5.2) during my interview with Afonso Mukai, one of the team working closely with Kafka.

Figure 5.2 attempts to unpack what happens to the data collected during an experiment in the time and space between the neutron hitting the detector and the information it represents appearing on the computer screen of the interested scientist. In Figure 5.2, data moves from the left to the right. On the left are three small rectangular boxes marked 'Detector', 'EPICS' and 'sample environment'. 'EPICS' is the control system that handles day to day operations of the experiments. 'Sample environment' is the environmental conditions surrounding the sample being tested. All of these (and many others that provide data on experimental conditions, but not included on my diagram) feed data into Kafka (the cloud-like bubble in the centre left of the diagram and the box next to it marked 'Kafka Node'). Here is Tobias Richter, Group Leader for Data Management talking about what Kafka does:

> You send data to it. It's processed detector data which means it's just the raw data, the raw information that you want from the detector. The detector delivers more information, or when capturing a single neutron, you get a number of signals. So we get a number of signals but we just want that to be refined to-, we had a neutron here, not, there were a number of sensors that detected a neutron and then you triangulate, or something similar. (Interview with Tobias, 2016)

Tobias is keen to stress that Kafka does not change the data in any way, but rather collects and holds the data until required by other programmes:

> We have basically just refined it to the bare minimum of the raw data. It doesn't, we don't know the wavelength of the neutron that can, that needs to be calculated later depending on other parameters that are then also in Kafka. So everything gets stuck into Kafka and Kafka just allows other programmes then to subscribe to that information. (Interview with Tobias, 2016)

Figure 5.2: Sketch of the data flow made by the author

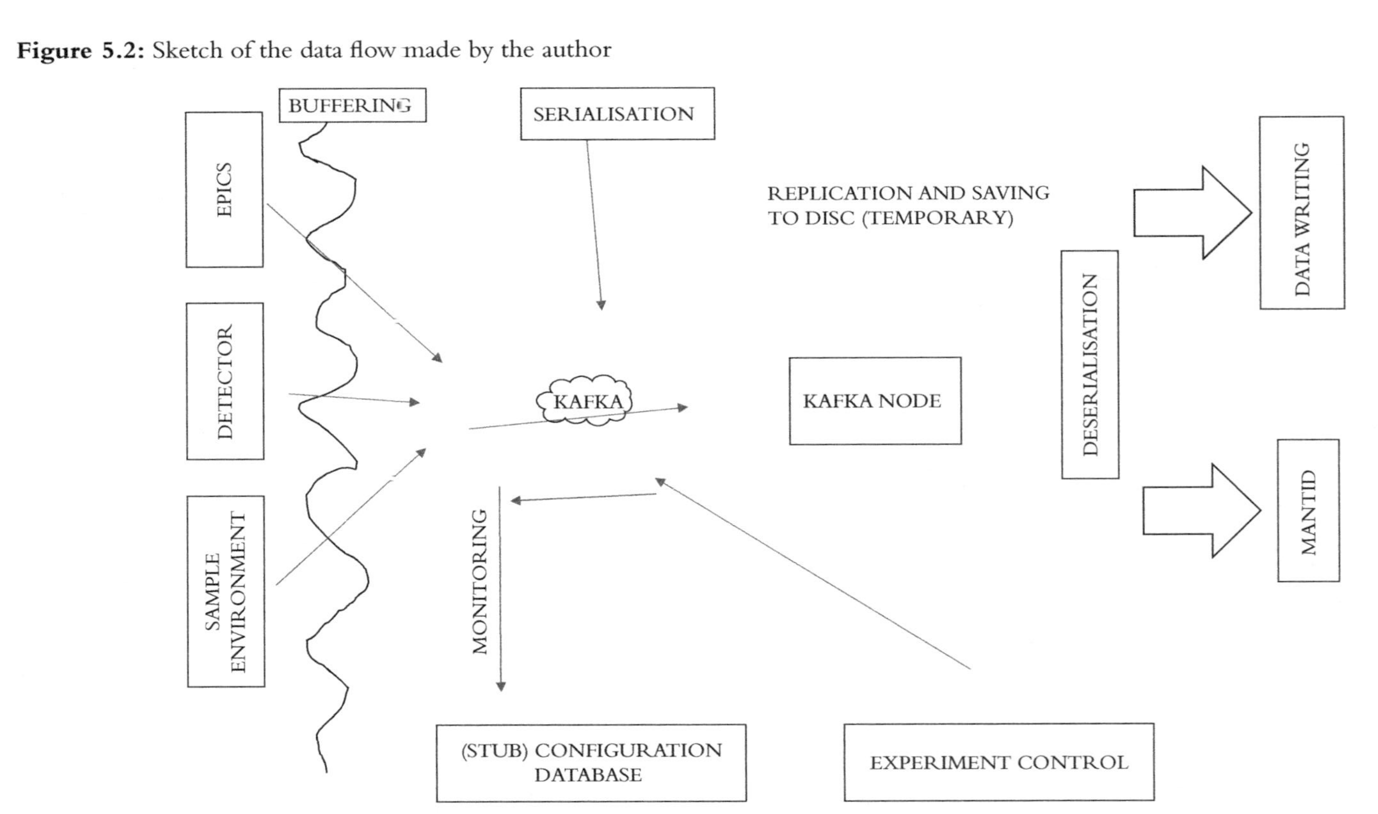

Note: This is a redrawn version of a hand-drawn sketch jotted down by the author during interview with Afonso Mukai. The sketch represents a simplified version of the flow of data from the detector.

On the right-hand side of Figure 5.2 are two small rectangular boxes marked 'Data Writing' and 'Mantid'.[2] Data Writing is the long term storage of the data gained from experiments. Mantid is an open source software framework for data reduction.[3] In order for the data to move from one 'box' to another, for example from the detector to Kafka or from Kafka to Mantid, it must be (what I call in the extract discussed later) 'disassembled' into bytes in order to be stored or transmitted, and then reassembled later. Formally these processes are known as 'serialisation' and 'deserialisation'. Here I am talking with engineer Afonso trying to understand this process:

Katherine: Maybe a better metaphor is something like you start off here with an IKEA sofa and you have to disassemble your IKEA sofa in order to transport it over here and then you get there and you have to be able to put your IKEA sofa back together.

Afonso: Yes, that's a really good metaphor, yes.

Katherine: I often find when I'm taking my IKEA sofa apart and moving it that bits get lost. ... Can the same thing happen here?

Afonso: Yes it could happen. For example if the network can't sustain the rates we're trying to achieve, it could start dropping some of these pieces, and that's one of the worries we have and one of the things we test the software for, so that we know that we build it in a way that it won't drop the bits, so that we can control it and avoid this. (Interview with Afonso, 2016)

Figure 5.2 also shows the three ways in which the DMSC take care not to lose data: buffering (in which data is collected and forwarded from the detector to Kafka in larger packets), replication (in which the data is saved to multiple places) and monitoring (in which packet size and volume is monitored to ensure they are not reaching levels recognised to be problematic for maintaining reliable service).[4]

Kafka is at the centre of a complex network of hardware and software that is designed to ensure reliable capture and movement of data from experiment

[2] Note that there could be many more 'boxes' on the right-hand side which receive data from Kafka in order to process it in various ways, for example, visualisation, simulation.

[3] Mantid was originally open source software developed on Github and evolved through community engagement and interest. Github was one of the best-known code-sharing sites, and its acquisition by Microsoft in 2018 was met with fierce opposition by some developers.

[4] It is also important to note here that experiments are often run multiple times so that results may be compared. This helps to mitigate further against concerns about data loss.

to scientist. The origins of the various pieces of hardware and software include both open source and commercial providers, with some developed in-house and some through collaboration/In-Kind Contribution (IKC). The development route for each is unique, drawing on discussions between the team members and their collaborators at other facilities and in industry. Once the individual pieces have been agreed upon, the technical experts must find solutions to join up the pieces and ensure a seamless fit and no loss of data.

Having now described the role of Kafka in the data management system, which is heavily focused on moving and storing data, I present two themes that emerged in the conversations about Kafka.

Relationality

Ensuring that the different components shown in Figure 5.2 are aligned (so that the data can flow smoothly) is all about setting up 'good relations'. This could mean making sure that the different technical components have been configured so that they can communicate with one another, or that the various human technical experts are connected in ways that facilitate sharing of expertise and experience. This theme comes across clearly in the following extract from my conversation with Afonso where he is describing what his role entails:

> Apache Kafka is an open source software project and it's available as a finished product. We're not developing Kafka itself, but what we do is, ok so we have to install Kafka to some machines and configure it, then we have to er install some dependencies, some other software Kafka depends on, and then what we actually develop is those pieces that sit close to the detector for example and take the detector data and feed it into Kafka. So and what I've been working I think mostly with recently is I'm setting up this infrastructure and getting for example, we have this piece that gets data from detectors and sends it into Kafka. We have some software which was done by our in kind partners and what I was doing is getting this to talk to our Kafka installation. And now for example, I'm deploying some other simulation that's from er our partners at PSI that simulate some components which are in the instrument so that we can have them there and then feed the simulator data into Kafka to see if the system works as a whole. So now I think my main work, I think the last months was setting up this kind of test infrastructure to see if everything integrates well. (Interview with Afonso, 2016)

I have shared Afonso's description at length because it contains many clues as to how he works with Kafka. In particular, Afonso's outline of his role uses many terms of relationality: 'dependencies', 'sit close to', 'feed it into', 'sends it into', 'getting this to talk to', 'feed the simulator data into'.

His work focuses on ensuring that the boxes marked on Figure 5.2 are not self-contained black boxes, as my diagram suggests, but rather more porous entities capable of taking in and passing on data as part of a flow from detector to scientist. Afonso's outline above also shows the different human relationships he must align in his work: 'open source software project', 'in kind partners', 'partners at PSI'. Unlike other Big Science facilities – including Afonso's previous employer in Brazil – software is not being developed in-house at the ESS but rather delivered as a process of contributions that represent different countries investment in the facility:

> we did a lot of stuff in house in Brazil, also because many things were not available through the local industry, so I think a lot of the expertise was available in house and in that aspect it's quite different here I think. I think here in usual big science facility projects, a lot of things come from industry and I think specially in the case of the ESS we have this spread development and that's very different from what I was used to. (Interview with Afonso, 2016)

Afonso's previous experience shows some of the specificity around the development of the ESS. As a major European project, it demands high level cooperation but also can draw on a large pool of expertise from across all the contributing countries. This is part of what will make the ESS a distinctive, cutting-edge facility. For those who must actually turn the ESS vision into a reality, this means employing people who are able to form good working relations between different technologies and different groups of people. Afonso is one of those who must make all of these contributions fit together, meaning that personal and technical relations are tightly interwoven:

> Right now what has been done is just different pieces have been developed by different people, so at different places, yes? And I've been, I think my work so far has been more of integrating things and the development I did was more into this configuration database. So but a lot of the work I've done here is actually integrating these pieces that come from different places, and putting them together. (Interview with Afonso, 2016)

With Afonso's help I drew the diagram that became Figure 5.2. Listening to our interview again, I focused on the arrows and wiggly lines we had drawn to connect the big boxes with proper names such as Kafka or Mantid, or that marked processes such as buffering of data. The boxes are either bought in from commercial concerns, or provided by in-kind contributors or other third parties. The lines and arrows between them were the spaces of relationships, where things feed from/into, integrate, indicate proximity or dependency between technologies. The construction of these arrows and lines is brought to life in my conversation with Afonso, when he talks about the collaborative aspect of this work. Here he describes the process they went through with one of their close collaborators at another facility (PSI) to start defining system requirements:

> We don't have a very formal process but I think it happened mostly at these meetings we had in person at PSI where we sat down and started to think, ok what kind of data we will have to deal with. Then we start listing and I think a lot of it comes from the previous experience of some of us who worked at these kinds of facilities before. So for example, Tobias, he worked as, with scientific computing and a user facility. And also people like PSI, they have a working facility there in ISIS so do have experience with, ok what kind of data are you going to send. (Interview with Afonso, 2016)

Working across organisations and countries, and with multiple colleagues, Tobias and Afonso work to align diverse expertise and technologies starting by forming consensus on the data itself (Gitelman, 2013). These are the 'seams' that join together the various pieces of software and hardware. They are designed to hold, to be permanent and stable *in order to guarantee the reliability and reproducibility of the experiments.*

Relationality is also a function of Kafka itself. Kafka is a little like a train station or transport hub, where data arrives (from the detector or control system, for example), waits for its next journey and then departs for processing or archiving. In this way, Kafka holds together the system, functioning as a hub that connects the many different components. Relationality is therefore something that is performed both by people and by machines in this setup. A reliable experience is dependent upon all technologies being well aligned. In order to achieve this, Afonso and Tobias must also work to align different contributors to the project. They must ensure components can 'talk' to each other, that the involved humans can collaborate effectively and that there is provision for malfunctions in the systems. The node structure of Kafka has been explicitly designed to avoid 'friction' or down time as it comprises many different individual machines connected together. In case of an upgrade or fault, an individual machine can be swapped out while the node remains functional. This is important so that the flow of experimental data is reliable.

In Figure 5.2 I sketched a very simple outline of what happens to the data while an experiment is taking place. There are two things to stress here. First, the diagram I drew is a gross simplification of a massively complex process with many more boxes (made up of both software and hardware) than are shown here. Many of these (black) boxes have their own development histories before and outside the ESS, which fade into the background when they are deployed by the DMSC. Second, when the ESS is complete the data will move through this sequence of boxes in real time. Meaning that scientists will be able to make adjustments to their experiments 'on the fly'. The transpositions and processing of data described in Figure 5.2 (serialisation and deserialisation) which are essential for aligning the different pieces of kit are mostly invisible to visiting scientists. The interviews contextualise each of the boxes in terms of it being produced by a particular contributor or as the result of a particular set of experiences. They make visible the network of human relations that stretch out from Kafka such that we cannot avoid noticing the 'contingent and contested social practices' (Dalton and Thatcher, 2014) that shape the data management taking shape at the ESS. The diagram I drew with Afonso and his descriptions of his work make visible the daily work of joining up people and machines, work that is invisible to users and the wider organisation. Afonso and Tobias' accounts also allow us to see how entangled the relations between people and technologies are, and to get a closer look at the actual workings of each piece of kit. This close-up provides different examples of what I have called in the next section 'tidying up'. It includes but is not limited to reduction of data, precautionary measures put in place to guard against loss or disruption of data flow, and changes to what data is collected.

Tidying up

> In order to serve their evidential function, data need to be adapted to the various forms of storage, dissemination and re-use over time and space to which they are subjected. (Leonelli and Tempini, 2020, 6)

The data looks different at different stages as it moves from left to right in my diagram. One of the ways in which movement between these different stages is facilitated is by its translation from coordinates in space at the detector into a series of numbers in Kafka. Alignment between different parts thus involves a kind of translation in order to smooth relations, a translation that also involves what I call in this section 'tidying up'. I chose this term to describe these processes because my participants view this as necessary work in order to achieve successful data journeys that don't fundamentally change the data. 'Tidying up' has a connotation of placing things in their correct place; making the message clearer rather than changing the message. It is connected to relationality because it is deemed necessary for the data to travel and be understood. Here is Tobias talking about it:

> I mean it's just raw information, there was a neutron here, there was a neutron there, right. And if you get repeatedly told where a neutron was, that doesn't, I mean even to a scientist in that field it doesn't mean

> anything, right. So you have to build up like an image to say, ok on average there's more hits here than out there, so it means I get some spot here. (Interview with Tobias, 2016)

By the time the data reaches Kafka some of it has already been tidied up, so that a clearer picture ('we had a neutron here') is available for the scientists. There is no clue as to what happens to the additional data, the white noise ('there were a number of sensors that detected a neutron and then you triangulate'), that is cleaned away. The 'cleaning process' has been performed in line with an assumed use/r in mind, but the exact characteristics of that use/r remain obscured (although hinted at in Afonso's comment that 'we sat down and started to think, of what kind of data we will have to deal with'). Tobias' comment is interesting because in a few sentences we see a difference between the data in the 'D' (detector) box and data in Kafka caused by processing which is designed to remove replications and produce a clearer signal, but which still leaves the data 'raw'.

Drawing the diagram with Afonso unpacks the steps in this early journey that the data takes. The promise of 'real-time data streaming' may give the impression to potential users of the facility that the data collected by the detectors is provided directly to their computer screens. However, the simple act of unpacking how this is done, the drawing of many boxes, lines and arrows shows the various steps comprising this journey. Unpacking the various steps that must be taken also makes space to ask questions about what happens at each step, and calls into question the 'rawness' of the data. Listening to Afonso describe the work he must do to join each step together shows that these pieces do not just fit together organically but rather must be made to do so, that there is labour, negotiation and skill here in doing this. It also lifts up the mutability of the data, showing how it is 'tidied up' for the technologies, to ensure a smooth flow through the system.

> Afonso: they all have to read this data from Kafka. For that they will have to use some special piece of software that understands what Kafka does, and inside this box, when it will get this data from Kafka and then deserialise it into a format that these programmes can understand.
>
> Katherine: So the detector, Kafka and then these boxes at the end, all speak different languages effectively. (Interview with Afonso and Katherine, 2016)

Thus, the data is adapted not just for different human practitioners, but also for different parts of the data management system. This newly 'raw' data is then deserialised in order to be transposed into Kafka – a process which Afonso admits may result in loss of data but which they try hard to mitigate against.

In 'Data Cleaners for Pristine Datasets?: Visibility and Invisibility of Data Processors in Social Science', Jean-Christophe Plantin discusses the processes of cleaning data for inclusion in an archive, and draws a parallel with the creation of scientific facts. In both cases, he argues, all traces of the processing are removed in order to produce 'valid' scientific knowledge and to give the appearance of objectivity. While the case he discusses refers to much later in the data life cycle than the one under consideration here, the connection he makes between 'pristineness' and 'a misleading conception of data as "raw"' also holds true here (2019, 55, see also Denis and Goëta, 2014).

In both Tobias' and Afonso's accounts there seems to be a widespread notion that the processes and technologies they describe are transparent, that the data passes through them without any effect on the message, for example Tobias talking about Kafka: 'It just sends on what it received. It doesn't change it at all. So it just allows you to say, I want … if you see something coming on that topic, I'd like to have it and make sure you get this.' However, here and there we can see moments of slippage when this understanding has to be clarified or confirmed, and where questions might be posed. For example, Tobias' comments that: 'It's processed detector data which means it's just the raw data, the raw information that you want from the detector.' Reading his words I am curious about how his understanding of 'processed' differs from my own. It is clear that there is some level of data processing that takes place which – in the understanding of the technical experts – does not affect the data. This processing happens between machines and with no specific scientific question explicitly attached to it. It therefore seems to take place outside the scope of the experiment. Placing Tobias' comment into dialogue with Bowker's assertion that 'raw data is an oxymoron' reveals that for Big Science experiments, the human contact is what starts the process of 'cooking' data (with 'cooking' used here as to mean processing, rather than anything fraudulent). If the data has just been touched by machines then it seems to be – at least in the understanding of the DMSC workers – still 'raw'. Only when it reaches the scientists who analyse it do we enter the realms of 'cooked' data. Bowker's provocation allows us to step back through the process and examine critically the journey that the data undergoes from much earlier, through the choice of hardware, software, the way in which it is serialised or deserialised, or the ways in which the technologies are aligned. But also potentially to return to the set-up of sample and detector which presupposes certain information being collected. Bowker's assertion works well here as a tool to open up a part of the experimental process that typically is hidden from view, the data journey from detector to scientist.

Before data from the ESS detectors reach the scientists conducting the experiment, they undergo various forms of adaptation in order to journey through the different components (serialisation and deserialisation) and also to give a clearer picture of the results (triangulation). Both of these are considered in this section under the general heading of 'tidying up'. The term was chosen quite deliberately as a gentle provocation to the ethos of the work done by my participants. 'Tidying up' perhaps implies that some unwanted or untidy data is removed, or put in its 'right' place (wherever that may be, and whoever decides upon that place). It engages with the idea that the alignment work being done here shapes the production of knowledge, while at the same time acknowledging that this work remains somehow outside the scope of the experiment, more or less invisible to the scientists themselves so that the 'tidying' does not compromise the 'rawness' of the data.

Making all the bits work together: data journeys and data friction

How can we understand this complicated journey that the data takes from detector to scientist? What does it give us to dig into the different steps and technologies? As I hope the previous section made clear, the data are not unaffected by the journey. Previously, I have focused on the themes of relationality and tidying-up as a way to put critical perspectives on 'raw' data into (hopefully) respectful conversation with the production of 'valid' data that is a prerequisite for a successful experiment. In what follows I will develop this critical perspective further by drawing on 'data journeys' and 'data friction' to make clearer how understandings and subsequent production of data are situated in highly specific contexts. 'Data journeys' (Leonelli) and 'data friction' (Edwards) are two concepts which stress the problems, glitches and difficulties involved in moving data and they are provocative conversation partners to my participants' accounts of the 'raw' data. Here I use these concepts to deepen my exploration of aligning different parts of the system that Jon lifted up as the big challenge for the DMSC.

Let's start with alignment …

In my analysis of the materials I collected, I wanted a way to focus on what happens to the data when it transitions from one of the 'boxes' in my diagram to the next, and to examine the seams between the boxes that require stitching together and smoothing over. Each of the components has its own distinctive development trajectory resulting in a heterogeneous collection of technologies that must be connected in a way that ensures a successful journey for the data from detector to scientist. Here alignment work provided valuable inspiration in thinking about how to handle this heterogeneity of software and hardware (Vertesi, 2014).

Alignment provides a focus on the moments of joining, on the seams or the work that must be done to achieve good connections between disparate artefacts with the goal of conveying coherent information. Janet Vertesi's account of alignment work tends to focus on the 'fleeting moments of alignment suited to particular tasks with materials ready-to-hand' (2014, 268) in a way that lifts up what might be otherwise neglected instances of invisible work. I take inspiration here from Vertesi's attention to the seams that exist between different technologies, and develop it as a way to focus on attempts made at the DMSC to create long-standing, stable, reliable seams. The notion of the 'data journey' usefully complements the focus on alignment by lifting up the directionality involved as the data moves from sample to scientist, navigating in more or less bumpy ways the transition from one component to another. The seams under consideration here occur at the very early stages of the journey made by experimental data. Involving the initial capture of data, they involve both hardware (from servers to pipes) and software, and take place in the geographically limited space of the Lund-Copenhagen area at the border of Sweden and Denmark.

Using the framework of the 'data journey' to understand the alignment work around data management at the ESS allows several important insights to emerge. First, it highlights the different stages in the lifecycle of data. The focus here, for example, is on what Leonelli calls 'data birth' as new ('raw') data is produced through experimentation, rather than re-use or sharing of existing data. The priority is on reliably capturing large quantities of data in real time as they are generated from one-off experiments. Thus 'alignment' here refers to the work done in a very small part of the data life cycle (so small temporally that it takes place in real time). A part that will be – for most users at least – invisible, albeit essential.

Second, 'data journey' as Bates et al detail in their account of the development of the term situates this work by paying attention to practices of knowledge production 'in relation to the wider socio-material contexts and power dynamics shaping their development' (2016, 4). The ESS, and the technical community surrounding similar facilities, is – as already noted by STS scholars – highly distinctive and replete with assumptions/expectations about data. The IKC model itself, means that experiences from other facilities are integrated into the ESS development. Furthermore, some of the DMSC staff have explicit responsibilities to liaise with scientific user groups to feed into disciplinary requirements/expectations about data.

Finally, the notion of 'data journey' requires a close attention to how data is located in physical space, and with that 'the disjointed breaks, pauses, start points, end points – and 'friction' (Edwards, 2010, 2011) – that occur as data move, via different forms of 'transportation'' (Bates et al, 2016, 4). These 'breaks' are both the product of transitions between heterogeneous components, but also connected to the mutability of the data as different users adapt it for different purposes and the decisions that have been made about how to capture and process data that help to situate the data.

Conclusions

Alignment work helpfully brings attention to the early stages of the journey that the data produced in the ESS experiments will take, rather than the polished final result (the 'scientific result' that is published). Focusing on the journey opens up for the possibility of other journeys, other ways to be, other results. It deliberately works against universal objectivity and stable facts. It is thus one way of 'situating' the knowledge being produced, by looking at the work that is done by humans and machines together. In doing so, it starts to challenge the boundary lines of the experiment, and contributes to changing notions of what counts as 'experimental practice' by highlighting the role played by technical experts in carrying out a successful experiment (defined as one that produces reliable data). Drawing on Janet Vertesi's work about alignment helped me to pose questions to the process my participants described, bringing their relationships and processes into focus as activities where humans work together to stitch the seams between technologies.

Alignment work is a way to lift up the invisible work done by the DMSC to align various people and machines to move data from detector to scientist: opening up the 'black box' of data management by looking at data in movement. The metaphors of journey and friction work here as a way to pay attention to this movement and the inevitable difficulties/risks involved in such movement. To do this, I focused on two themes. Relationality is connected to context, breaks and friction by examining the ways in which different parts of the data journey are linked together by people and machines. Tidying up, meanwhile, is a key aspect of mutability as the data is arranged into a particular form in order to facilitate its travels. Both foreground the work involved by my participants and how their expertise intersects with the management of data.

6

Technician or Scientist?

In the previous chapter, I highlighted the development of a key part of the data management set up at the European Spallation Source (ESS) by examining the adoption and integration of Kafka. Kafka is an interesting case because – as Afonso noted – it was an existing piece of equipment developed in the commercial sphere and then adapted for use in the academic/industrial space of Big Science. Kafka was one of the clues in my material that suggested a change in how data management in Big Science is being organised, more precisely a shift towards recognising the value of expertise and equipment developed in contexts outside Big Science. This shift is epitomised in the use of the term 'professional' which my participants used interchangeably to indicate commercial (as opposed to academic) development, and expert (as opposed to amateur) skills. In this chapter, I will explore the idea that the work of data management in Big Science is becoming increasingly professionalised, and the consequences of this.

Historically, software to process the data was done on an ad-hoc basis, usually by the postdoc of a professor leading an experiment. This 'home-made' software would be passed down among the professor's group of grad students, who would make adjustments 'on the fly'. This paradigm no longer fits as the data streams get larger, and more complex, and the work of data management increasingly requires expert technical skills. It is also a challenge because the Big Science facilities are increasingly used by researchers from a wide range of disciplines, many of whom with no background in computer science or data analysis. This means facilities must start to provide technical and computing support to users in more formalised ways. This includes provision of standardised software for instrument control or data analysis, but also creation of a visible team of facility staff tasked with developing technical solutions and supporting visiting users. One way to understand this is to frame the work of the Data Management and Software Centre (DMSC) as the emergence of a new 'professional' group of technical experts.

Steven Shapin famously claimed in 1989 that there 'does not exist a single study documenting and interpreting technician's work' (p556). Since then, there have been important additions to the scholarly literature that address this gap. In what follows, I will show that the line between scientific work and technical work has never been more blurry than at the ESS. In this chapter, I want to put this demand for increasing professionalisation into conversation with scholarship from Science and Technology Studies (STS) about the distinction between technicians and scientists. This literature is attuned to teasing out the differences between these roles in ways that bring attention to different kinds of expertise and in/visible work. While some of the work that my discussants carry out could be easily placed into the realm of the purely technical, could their emerging status as data professionals also be impacting their visibility as contributors to The Experiment? For example, when visiting scientists are assisted in data management by the DMSC staff, are these staff members contributing to the scientific results and will they be included on the list of authors for resulting publications? Is the boundary line of what counts as part of The Experiment changing, and is DMSC slowly edging into that space? In short, what difference does it make if the work of the Group Leaders and their colleagues is becoming professionalised?

In order to answer these questions, I will start by introducing you in more depth to my participants, analysing their accounts of their own career trajectories to examine how ideas about 'professional' work emerge. I am going to put these into conversation with the scholarly literature concerning the distinction between technicians and scientists, as well as considering the impact of data-intensive science on this distinction. Can the dividing line between 'technician' and 'scientist' be correlated with the boundaries of The Experiment proper? Who is allowed to make knowledge and who supports the knowledge production process? What are the consequences of this dividing line for individual careers, organisational priorities and visibility? In the last part of this chapter, I will focus in on the consequences of this increasing professionalisation. How do the participants themselves put the different models of knowledge production to use amid the rise of a new professional technical class in a new era of Big Science? And what difference does that make to how knowledge is produced in Big Science?

Participants' backgrounds

At the time I started my project, there were five Group Leaders and the Head of the DMSC on my interview 'shopping list'. The Group Leaders I interviewed had a range of different background experience prior to joining the DMSC. My instinct when starting the fieldwork was that these experiences would be important to understand how they approached the work of data management at the ESS and thus it formed the basis of my very first questions to them. One of the first interviews I did was with Jon Taylor (Group Leader for Instrument Data at that time), who described his background thus:

> My background is in neutron scattering. I did a PhD in neutron scattering at the ILL and then I did a Post Doc at Warwick University doing X-Ray scattering and then I got a job at ISIS, which is the UK's neutron source and I worked there as an instrument scientist from 2002 to 2014. (Interview with Jon, 2015)

Tobias Richter (Group Leader for Data Management, and who you met in the last chapter) has, like Jon, a background working at other Big Science facilities:

> I am a physicist by training. My PhD is in atomic physics, it's sort of a slightly esoteric field by now. I've done experiments with synchrotron radiation soft X-rays. So I've been, well during the PhD I already went into some of the more programming related aspects of that. So I then mainly have a background in data acquisition. I moved to the UK Synchrotron. (Interview with Tobias, 2015)

Both Jon and Tobias had worked primarily in what we might call 'academic' contexts, doing their PhDs and then working at other Big Science facilities. Both had worked at different facilities across Europe, and extensively in the UK where there are two facilities located on the same site: Diamond Light Source (what Tobias refers to as the 'UK synchrotron') and ISIS. This was similar to the narrative of Mark Hagen, who was Head of DMSC when I started doing fieldwork:

> I've worked in neutron scattering for 34, 35 years or whatever, that's what I did my PhD in. It's fairly standard, you find people who've done these things, who've stayed in their PhD subject all their scientific career. You know, there are others who've done other things, but you know, a lot of other people do. So after I did my PhD in Scotland actually, up in Edinburgh. Then I went to work briefly as a Post Doc at the Institut Laue-Langevin which is in Grenoble, France, which back then was the French British German neutron scattering centre, it's a reactor. Then I was a Post Doc again in the United States at Oak Ridge National Lab, where the Americans have one of their two reactor neutron sources. They have what's called the High Flux Isotope Reactor. And I was there for a couple of years, then I came back to the UK and worked at the ISIS Spallation Neutron Source. (Interview with Mark, 2016)

Jon, Tobias and Mark all share experience working at the same UK site: Rutherford Appleton Laboratory (RAL), which is home to a cluster of laboratories all working with cutting-edge science, including Diamond and ISIS. RAL is managed by the Science and Technology Facilities Council, a key provider of infrastructure and funding for researchers in the UK. The site has existed since 1957, and is well-known in the Big Science community. So, while there are differences between the experiences of Jon, Tobias and Mark, I want to suggest that the shared experience of sites such as RAL is important in providing a common reference point in terms of infrastructure, users and technologies. These shared reference points also contribute to the informal network of contacts and knowledge sharing that would prove to be essential to the work being done by the Group Leaders.

In contrast, two other Group Leaders, Thomas Holm Rod and Sune Rastad Bahn, had more commercial experience. Sune, after his PhD, had extensive experience working with major companies including IBM and Microsoft. His account of his working life before ESS highlights very different aspects to that of Jon and Tobias:

> After that I was invited to join a start up that did some of these nanotech simulation software, and selling it into the industry to some of the larger electronic manufacturers … and there I started managing people. I grew up a division of software development and we ended up being something like 15, 20 people in my group. And it was very, it was an eye opener for me in terms of doing people management and stuff. I didn't have any background but I picked up a lot of things, read a lot of literature about it and, you know went to courses. And found out that it was actually a really interesting subject. And found, I was reasonably successful in creating a good team and make them effective. Anyway at some point out the venture capital somehow ran out and the company somehow folded. So then I moved on to IBM, was working there as an IT architect for some years. (Interview with Sune, 2015)

Sune found out about the position at DMSC that he held through Thomas, with whom he had worked previously. Here's Thomas (2015) telling his story of how he came to the DMSC:

> so I'm a research scientist myself, so PhD, actually same place as Sune. Then I have three years post doc experience from US, scientific software, mostly as a user rather than developing it. I've developed some but not professional software development alone. Two years of post doc from Lund University. Then I went to a company where Sune also worked, a startup company. (Interview with Thomas, 2015)

When the company folded, Thomas tried starting his own company but with no success. Looking around for work, he spotted the job with the ESS and was involved in writing the technical design report for DMSC, making him the longest serving member of the management team when I first interviewed him. While Jon, Tobias and Mark pepper their narratives with names of universities and facilities, describing their various roles as 'PhD', 'postdoc' and 'instrument scientist'. Sune's narrative focuses on the shift in his skills into leadership and management. Thomas frames himself as 'a research scientist' and 'a user' of software 'rather than developing it', as well as, later, an entrepreneur. Together, this group represented the technical management and leadership of the DMSC at the start of this project. All have PhDs, but the ways in which they developed that research experience varied, spanning academic to commercial contexts. In contrast, the fifth Group Leader, Petra Aulin, has quite a different background:

> I'm a Swede and I moved to Denmark for studying, and I studied at Copenhagen University, Regional Development … and when I was done I started to work at the Øresund Bridge, the company, not on the Bridge [LAUGHS], but on the company, with regional development questions. How do we get Swedish people to use Denmark and vice versa and to improve and to try to make the cross border problems easier to, yeh easier access to the region. (Interview with Petra, 2015)

Petra's role during the time that I was meeting the Group Leaders was focused on managing the challenges of a having a cross-border organisation. Her role was to negotiate with the Lund ESS site, handle budgeting, specify the new office premises for DMSC. As such, she was at the heart of the DMSC organisation and will feature much more in the following chapter about organisation.

The career trajectory narratives of Jon, Tobias, Mark, Sune and Thomas gave me my first clue as to how the technical work of data management might be changing in Big Science. More precisely, their backgrounds seem to suggest two pathways into their work at the DMSC – one more grounded in commercial software development and one emerging from academia/close work with the facilities. Sune and Thomas' narratives bring an emphasis on professional project management and people skills into the conversation. During our conversations, they reflected more explicitly about *how* to do the work, educating me on various programs to manage tasks such as Slack, and reflecting on the different organisational cultures of which they had experience. Jon and Tobias' accounts highlighted how research backgrounds might easily evolve into software development in the case of Big Science, providing them with many contacts in the field and direct experience with users. During the conversations with the Group Leaders, they were also the people who reflected less on the organisation as a whole or the management style. It was also possible to trace some preference in terms of technical solutions; a commercial background had given Thomas and Sune experience with a wider range of hardware solutions with no pre-existing preference for solution, as well as professional training in how to handle a team. Jon and Tobias, however, were familiar with the in-house development approach that characterised Big Science software for many years. These differences were most clear in the first year of interviews.

What did those different backgrounds mean for their work at the DMSC? How might this impact data management at the ESS? I want to suggest that the differences that I saw in their narratives are – like Kafka – indicative of a broader change that is happening in data management in Big Science. This change is marked by the increasing influence of 'professional' approaches to data management, characterised by formal development of management skills, experience in the commercial sector and willingness to consider commercial solutions. What is understood by the term 'professional' became an interesting focus in our conversations as the Group Leaders used it to reflect on their own work, as well as the organisation more broadly. Here, for example, is Jon talking about 'modern software development' in relation to the work being done at DMSC: 'What they really mean is that we've taken some principles and we've applied them from the professional domain. Now OK that's assuming that the professionals get it right, which is a question mark. And we are somehow professionals as well' (Jon, 2016).

Jon is here describing methods of testing software, but in the middle of his description of the process he distinguishes the work they are doing from that of the 'professional domain' indicating that their work is somehow more academic perhaps. As he goes on, however, the demarcation of who and what counts as 'professional' can be seen to be increasingly blurry: 'we are *somehow* professionals as well' (my emphasis). Meanwhile, here is Sune talking about changes to the planned development of the ESS (particularly in relation to the rebudgeting exercise that occurred around the time of this interview) and professionalism: 'it stands out to me that all these changes have not been done in a what I would consider professional way. I mean, they are, you know you can read books about change management. It's not like nobody tried that before' (Sune, 2016).

During the previous year the organisation as a whole had undergone several changes and – in this quotation – Sune is reflecting upon how that change was communicated and implemented in the organisation. Both Jon and Sune's accounts therefore point to the ESS as not following processes that they feel are common practice in commercial organisation (here 'professional' in their accounts seems to indicate commercial organisations). It is also notable that in Sune's account doing things in a 'professional way' seems to have a positive connotation, while Jon's use of the term is much more ambivalent.

In this section, I have illustrated some ways in which the Group Leaders at the DMSC distinguish the work they are doing from that of professional software development, and instead align it more closely with an 'academic' model more akin to the in-house development taking place at many Big Science facilities. The use of 'professional' across a number of the interviews seem to indicate a hierarchy of expertise to which my discussants compared their work at the ESS. It also made me wonder about the changing status of the Group Leaders themselves.

I selected the earlier quotations because they illustrate how the notion of being 'professional' functions as some kind of level of quality to which the Group Leaders aspire. It covers both software development and people/organisation management. In all cases, the Group Leaders frame ESS/DMSC as somehow falling short of 'professional', and themselves as introducing 'professional' behaviours or standards: 'we are somehow professionals as well'. These comments and reflections stand out in contrast to previous accounts of how software was developed in Big Science.

Data management is no longer a service relegated to an invisible, underfunded small group of people at the facility (Seidel, 2008), with the majority of the reduction and analysis work actually being performed by skilled users. Instead, the DMSC represents a visible organisational unit with spending power sufficient to buy in a piece of equipment such as Kafka. It is staffed by a mixture of people, with both academic science backgrounds and commercial technical backgrounds. It feels like something is shifting – the work of data management is in and of itself becoming more (as my participants termed it) 'professional'. There are a number of threads to pull here to follow this discussion a bit further. One concerns standardisation as a marker of professionalism, more precisely the move from individual researchers developing idiosyncratic solutions to handling their data to a facility wide, or even cross-facility consensus on interfaces and software to ensure a smoother experience for users who nowadays visit many different facilities. Another thread would be the changing status of the Group Leaders themselves in the experimental procedure. With their expert technical skills, advanced understanding of the science itself, and experience from both inside and outside Big Science, it becomes hard to see them as simply the technicians who make the data flow work.

Technicians, scientists or something else?

In what follows I will draw on the scholarship within STS that discusses how the roles of 'technicians' are understood, as compared with those of 'scientists'. This is not because I inherently agree that the work done at the DMSC is purely 'technical' in nature, rather that this body of literature is helpful to explore how the job of data management in Big Science is changing, what skills are valued/necessary in order to do this job and whether this is impacting the visibility of units like the DMSC.

> When the scientists do data analysis, it's part of their science and a lot of that is also included in writing or developing algorithms or code that enable you to do data analysis. So there is a challenge or there is a kind of paradigm shift, I guess you'd call it, when you say the facility is going to then do data analysis in a supportable maintainable way. (Interview with Jon, 2017)

The Group Leaders know very well that the remit of the DMSC represents a change to 'traditional' data management practices in Big Science. Jon, however, identifies the broader significance of this shift when he talks about it being 'part of their science'. I am interested in the relationship between in/visibility of data management practices and ownership of these practices. Is data management more invisible because it is considered more 'technical' than 'scientific'? And how does the shift from user to facility in terms of being responsible for data management impact this visibility? If the facility takes responsibility for data analysis, for example, does that mean the DMSC becomes part of 'the science'? Does this mark a clear departure from their status as primarily technical experts?

Arguably the best known piece of STS literature that discusses the dividing line between technician and scientist is Steven Shapin's piece, 'The Invisible Technician', in which he argues:

> technicians have been almost wholly invisible to the historians and sociologists who study science: in the now-vast academic literature in the history and sociology of science there still does not exist a single study systematically documenting and interpreting technicians' work, past or present. (Shapin, 1989, 556)

Shapin's article on 'The Invisible Technician' is primarily an historical view focused on the laboratory of Robert Boyle, in which dozens of assistants laboured to make Boyle's scientific ideas into realities (1989). Having set the scene of a laboratory populated by dozens of busy assistants, labouring according to Boyle's wishes and unacknowledged in the scientific results, Shapin poses two questions. First, how to document the role of these assistants who have been removed from historical records of how the research was carried out, and second, why were they so removed? Shapin provides a number of suggestions to answer this latter question,[1] which naturally turns

[1] Donna Haraway critically engages with Shapin's reading and reveals how those who were allowed to witness science (and therefore produce knowledge) represented the most privileged position in society (white, male, upper class, educated, etc.) (1997).

to a discussion of the extent to which these conditions remain prevalent in contemporary scientific endeavour, concluding that 'the historical arrangements and sensibilities described here are not wholly irrelevant to understanding the modern situation' (Shapin, 1989, 562). Some 20 years after Steven Shapin's famous text, in a piece simply called 'Technicians', Rob Iliffe seems to agree when he writes that:

> despite the attention paid to scientists' practical skills, and to the material culture of the laboratory, technicians remain absent from virtually all sociological and historical accounts of scientific practice. In a sense, they have remained obstinately invisible because sociologists have transferred to scientists various features that have usually been held to characterize the work of technicians. (2008, 4)

Both Shapin and Iliffe are careful to provide sociohistorical context that shows how scientific practices have changed and thus to delimit claims about any neat distinction between 'scientist' and 'technician'. In his introduction, Iliffe teases out a number of aspects of working practices that might be used to distinguish technician's work from that of scientists, all the while stressing that these distinctions were extremely porous and open to exceptions. He includes, for example, the way in which technical work may be tied to one location, be spoken of in terms such as having a 'knack' for doing something (or equivalent metaphors that seem to point to some bodily affinity with the work, see also Doing, 2009), and involve in-depth knowledge of *both* equipment and the science itself. I like Iliffe's article because it explicitly brings the sociologists into the picture and asks what their role has been in rendering some tasks and some roles more visible than others. In doing so, this becomes a more meta-level reflection on the role of the observer-researcher in determining what 'counts' in an account of Big Science.

This kind of reflexivity is also evident in some accounts of Big Science, such as Park Doing's *Velvet Revolution at the Synchrotron* (2009) in which he narrates his own developmental trajectory working at the Cornell Synchrotron initially as an 'x-ray laboratory operator' and later as 'assistant operations manager':

> Many operators understood the tag of 'operator' as a taint, a stereotype imposed upon them by scientists who ignored their input into laboratory matters. Once you were cast in the role of operator, according to this line of reasoning, you were stripped of the credibility necessary to be a valid knowledge contributor in the laboratory. (Doing, 2009, 53)

So, in Doing's own initial experience as an operator, technicians are somehow shut out from the formal knowledge production process. However, as his own role developed to include more contact with the scientists at the Lab, his opinion changed: 'Neither the scientists nor the operators were "right". Rather, they were each promoting and performing their own point of view, their own idea of who can produce technical knowledge and how it is produced' (Doing, 2009, 64).

In coming to this realisation, Doing grounds his observations in STS literature attuned to studying different models of knowledge production. It is also an observation that resonates strongly with me. When I shared early drafts of these chapters with participants and other experts, the differences in how we each understood what constituted 'valid' or 'useful' knowledge were striking! The texts I wrote about the DMSC disappointed, for example, some of my participants who felt that I did not fully understand or appreciate their technical achievement. Meanwhile, the same texts produced bafflement from users of similar facilities who depended on the epistemological stability of the 'raw data' category for their experiments to be reliable. While the failure to communicate more clearly my epistemological standpoint and goals is entirely my fault, this experience also illustrates rather aptly how different models of knowledge production pay attention to different things, deploy different methods to capture different kinds of data, and consequently render different kinds of work more or less visible in the pursuit of science.

This point is developed further by Garforth in her article 'In/Visibilities of Research: Seeing and Knowing in STS' (2012). Here the question of what is visible or not is as much due to the gaze of the observing social scientist as it is to the participants themselves, and claims as to in/visibility should therefore be handled with care. Garforth draws attention to how much of the STS work that delves into making practices visible focuses on the element of the visual. The importance of a researcher actually observing 'science-in-action' is paramount in making a truth claim about that particular knowledge production context. How then to work with this in a context where much of what is to be studied is literally invisible because it is in the form of digital data and software? The emergence of what is sometimes called 'data-intensive' science introduces challenges to validation in the form of equipment not previously recognised as 'instrumentation' but also in the form of new tasks to be performed, as Scroggins and Pasquetto highlight:

> Data-intensive science has added a digital overlay to existing maintenance requirements; adding datasets, analytical pipelines, archives, and software to the list of scientific instruments to be maintained and calibrated. As well, the digitization of so much data has seen scientists and their apprentices take on ad hoc maintenance and repair tasks, primarily when increasingly complex analytical pipelines breakdown in the midst of analysis. (Scroggins and Pasquetto, 2020, 122)

Data management in Big Science is invisible to most users until it breaks down and disrupts the flow of the experiment. This work may also be less visible to researchers because it is literally harder to observe, or it may be overlooked because it has been categorised as 'technical' rather than 'scientific'. These forms of invisibility may place it more in the role of supporting an experiment rather than contributing to the results. However, what is clear is that any kind of distinction between technician and scientist may be more blurry or dynamic than it seems (Iliffe, 2008). Furthermore, the work of those with technical expertise is highly valued by users – valid just different (Doing, 2009). And finally, in lifting up the work of the Group Leaders my own account of their profession skews things – I see certain things and not others (Garforth, 2012). The inevitable question becomes: what then is the effect of increased visibility for data management practices?

Impact of increased visibility

> it is the first time I think that money actually has been allocated at a facility to work on that, and I would say it's always, I feel constantly at risk of being cut. (Interview with Thomas, 2015)

Thomas is Group Leader for Data Analysis, and during our first conversation he pointed out that this part of the services offered by DMSC was unusual for facilities. Previously the data analysis part of the data management pipeline has been carried out by the users themselves. Provision of data analysis services by the facility is thus one of the many ways in which data management is becoming more visible at the DMSC. Providing new services, like analysis, has an impact not only at the organisational level, but also in terms of which users might be attracted to using a facility and the kinds of expertise required by staff at the DMSC. However, as the earlier quotation illustrates, this new service occupies a tenuous position, constantly at risk of being cut from the budget as the most recent addition. This ambivalent, precarious position is, however, interesting to me in its very precarity because it is at moments like this when the remit of data management can be seen changing.

In Thomas' comment there is a very real sense that the scope of data management in Big Science is changing right now. The data management work has at the ESS been recognised as a separate organisational unit, the remit of the data management work has broadened compared to other facilities, the interviews with the Group Leaders show the influence of external/commercial experience and technologies, and an ongoing identity negotiation in terms of their own professional role. All of these changes are potentially significant in terms of visibility of data management and the impact on the work being done. In this section, I want to return to the reasons driving these changes as they are understood by the Group Leaders, and to ask what is at stake.

All of the Group Leaders acknowledged changes to the academic climate which are driving more demand for the services being prepared at the DMSC. These span the size of the data set, different kinds of users and increasing competition between users:

> So whereas it used to be that you know you had a few data and you wanted to do a lot of computation on those few data. In a lot of areas today … you have a lot of data and so there's a tendency that you move your computation to where your data is, rather than the other way round. (Interview with Sune, 2017)

> They're not career users anymore. A more career user is a person who does use neutrons for science and does that during their career, only that, so that's the focus area. Now it's more like one tool in the toolbox, right, so you go to a neutron facility, you go to do an experiment, you go to X-Ray, you use many different technologies and that means that users need more help. (Interview with Thomas, 2015)

In Sune and Thomas' reflections, they both contrast an earlier time ('it used to be that …', 'they're not career users anymore') with the contemporary situation, highlighting how this changes what users need from facilities. These are important external changes to the academic climate and could reasonably be assumed to be driving a greater demand from users for data management services at Big Science facilities like the ESS. Consequently, the responsibilities associated with DMSC are different to those in equivalent groups at other facilities and the Group Leaders reflected openly on that change. Increasing regulation of research data by funding bodies and journals also generates more work with data, while potentially changing the parameters of what is considered the work of doing science.

Publication is a visible, high stakes arena for validation and acknowledgement of scientific results and expertise. The Group Leaders publish papers in their own right, documenting the work they have done but also might become authors on papers of the experiment depending both on the extent of their contribution, and the willingness of the researcher to include them (Scroggins and Pasquetto, 2020). Preparing data sets in such a way that they accompany publication and fulfil the requirements for shared or open data is also widespread now. However, this work of preparation produces a new category of workers (both technical and administrative) who are experts in preparing data sets and working with accompanying infrastructure. Despite the necessity of such work, it is often a new class of 'invisible work' unrecognised in publications or funding (Scroggins and Pasquetto, 2020, 117).

From this angle, the changing practices of scientific research, particularly in relation to publication and data sharing seem to favour increasing validation and visibility of people like my participants, while simultaneously producing more invisible work for others. There are also clear demands from users for the kind of expertise that the Group Leaders and their teams provide. All of which would suggest an increased level of visibility around data management in Big Science, which we see resulting in concrete changes such as the decision to fund parts of the data management pipeline previously not considered within the remit of the facilities. It is therefore perhaps unsurprising to hear the term 'professional' cropping up so often in the conversations I had.

Conclusions

My participants' backgrounds reflect that both 'technical' and 'scientific' skills comprise important experience when it comes to building the DMSC. In STS scholarship this distinction has been used as a way to explore how different kinds of knowledge and know-how are valued. In the specific context of the ESS, the role of the DMSC Group Leaders epitomises broader shifts taking place within Big Science that suggest a changing understanding of the role of data. Here I have attempted to unpack the distinction between technician and scientist through the lens of visibility, using the discussants choice of the term 'professional' as a focal point. This has highlighted negotiations around professional identities of the DMSC staff but also the power balance in the work being done. An acknowledgement of a growing dependency by users tips the balance towards the staff with 'technical' expertise who become essential to carrying out An Experiment. In this moment, the boundary line around what and who 'count' as part of An Experiment shimmers and blurs in ways that allow me to ask questions about what is at stake at the border line of technician and scientist in Big Science.

As was clear in Shapin's account of Boyle's laboratory, technicians are invisible in the public account of scientific knowledge production. While technicians were essential to the actual practice of science, it was Boyle's name that was famous. Similarly, the staff of the DMSC may not always be included as co-authors on publications that result from experiments at the ESS, but their work is essential to the smooth production and management of data. This invisibility makes it harder to study their work and suggests less professional validation.

Unlike scientists, technicians can remain invisible and out of the experiment, thereby allowing them to work with the data and the data still to remain 'raw' when it reaches the screens of the visiting scientists. The downside to this is that data management and scientific computing has often been poorly funded in Big Science. However, if the staff at the DMSC become part of The Experiment Proper (meaning, for example, included in diagrams explaining how the ESS works, regularly appearing as co-authors on papers, funded in line with other parts of the organisation) then their professional status potentially changes, walking the precarious boundary line between technician and scientist. Becoming data management professionals in this way connects my participants to wider (read: outside academia) discussions about data and makes them more visible. This is not just a result of the changing requirements from facility users for more support in terms of managing their data. It is also connected to my participants' accounts of their own career trajectories. As we see with the DMSC Group Leaders, those working in data management at the ESS represent a variety of backgrounds, some more commercially oriented where recognition and validation of data management expertise is more visible through its professionalisation.

What impact, however, do these shifts have on the data? If a certain amount of invisibility (achieved in part through the use of the category of 'technician' to describe their work) helps to maintain the data as 'raw', then what might be the consequences be of increasing professionalisation? Does recognising the skills of data management experts such as my participants inevitably mean a recognition (and increased visibility) of the technologies they deploy to manage the data? Examining the tension between 'technical' and 'scientific' work in this context thus allows us to see one way in which the 'rawness' of data in Big Science has been preserved, what is at stake in doing so and what might change as the profession itself changes.

7

Organisational Frictions

During the first year of my interviews (2015), the Data Management and Software Centre (DMSC) was housed in a temporary home in one of the older University of Copenhagen buildings (Figures 7.1 and 7.2). This building was a rather worn-looking building, where one was greeted with the sight of student lockers and the smell of old socks when entering at the ground floor entrance. Up several flights of stairs and the DMSC occupied several smaller rooms grouped around a large kitchen. By the second year of interviews, the DMSC had moved to its 'permanent' home in a new commercial building, with a smart reception area, shared with various biotech and infotech start-ups, although still near the university (Figure 7.3). Here they sat in an open plan space with three meeting rooms at one end, and shared a kitchen with other businesses on the same floor.

This move from one premises to another came up in all of my conversations with the participants and touched on planning, work relations, and status of the DMSC within the wider European Spallation Source (ESS) organisation. In some ways it seemed to mark a new chapter in the DMSC journey, as they moved to a space that addressed long-term planning needs such as staff recruitment and resources for a server room.

In the previous two chapters I have focused on technical affordances and personal expertise, framed through the specific examples of the alignment work around Kafka and professionalisation of technical work within Big Science. In this chapter I pay attention to the organisational context in which the work of the DMSC takes place. What can studying the organisational practices and culture of the DMSC tell us about data management? This chapter is premised on the idea that studying the processes, procedures and documentation of an organisation provides clues as to why some kinds of work are visible and others not (Garfinkel and Bittner, 1967; Hanseth and Monteiro, 1997; Star, 1999; Star and Strauss, 1999).

My instinct when I started this project was that a distinctive organisational culture would play an important role in shaping the technical solutions being developed at the ESS. However, the organisational culture itself

Figure 7.1: Entrance hall at previous DMSC premises at University of Copenhagen

Source: Author's own image

Figure 7.2: Ground floor entrance to previous DMSC premises at University of Copenhagen

Source: Author's own image

was in its infancy and during the three years of my fieldwork I saw many changes. In turning to look at ESS as an organisation, then, I take note of organisational resources such as premises and budgets, but also changing relations between people, groups, buildings and networks of contributors. As I hinted at in Chapter 5 in my discussion of relationality, these organisational aspects are entangled with both technical limitations and personal expertise, and play an important role in determining which technical solutions are selected, how they are developed and the success (or not) in implementing these solutions. In order to try and show these

Figure 7.3: Exterior image of the current DMSC offices, located in the COBIS building, in Copenhagen

Source: Press image courtesy of Symbion and BII

entanglements more clearly, in this chapter I will zoom in on three examples: the DMSC move to new premises, the In-Kind Contributions (IKCs) and the budget. These examples were selected by virtue of my participants' own enthusiasm to discuss them at length during our conversations. To help me dig into these examples, I will be returning to the earlier discussion about infrastructuring and invisible work (see Chapter 4) as a critical lens for exploring the relationship between technology and organisational context.

Infrastructures (both digital and material) are the scaffolding put in place to support organisational relations. The literature on infrastructuring has highlighted the difficulty of studying such relations for the reason that when they are working well, they tend to 'disappear' into the background. Data management infrastructures are as much subject to these as other organisational infrastructures, making the study of infrastructuring practices used by organisations to set up and discuss data management infrastructures perhaps doubly difficult to 'see'! However, during my time following the development of the ESS the organisation was in a high level of flux, which meant that infrastructures were being put in place, renegotiated and changed in ways that made these practices a topic of discussion and thus more visible within the organisation and to me as a visiting researcher. Here, for example, is Sune talking about organisational culture at the ESS during our first conversation in 2015:

> I would say there's a good deal of this startup feeling, because it's very much a project where there's a lot of open stuff where you can influence. There's a lot of decisions that haven't been taken. It's far from a train which is just running on a track. It's much like, there's much more exploration going on, people going in different directions, and

> there's a lot of, you know need for also creating structure where there is no structure. (Interview with Sune, 2015)

His choice of words conveys a clear sense of an organisational culture in early development with much opportunity but also plenty of scope for misunderstandings and confusion ('people going in different directions'). It is also an important reflection in that it shows how a lack of infrastructure may be more visible than an established infrastructure, and that an important part of the work of infrastructuring involves 'a lot of decisions'. These decisions constitute moments at which organisational culture as well as personal experience materialise into specific practices, or in the case of the DMSC into pieces of hardware and software. In this chapter then, I am paying particular attention to the moments in my interviews when the negotiations around decision-making are clear, and to approach these as moments where infrastructuring practices might be more clearly seen and analysed in terms of what they can tell us about the organisation.

In the case of the ESS, paying attention to organisational dynamics and politics involved in infrastructuring practices requires looking both 'inwards' and 'outwards' from the DMSC. By this I mean that the DMSC is shaped by both its internal relations to the rest of the ESS, and also very much by its external relations through the wider organisational network of IKCs whose expertise, ideas and contributions shape the ESS. In what follows, I provide a brief overview of the structure of the ESS organisation, before turning to the three examples mentioned earlier: the DMSC premises, the budget and the IKCs. Here I will include the voices of my participants and analyse their comments for what we can learn about how data management infrastructures evolve in dialogue with organisational culture and priorities.

Internal ESS organisation and structure

The ESS is distinctive in that it comprises two sites, separated by the Öresund channel that flows between Denmark and Sweden (see Figure 7.4). From its inception, the DMSC was planned as a separate site, located some 70 km away from the Lund ESS site (what is referred to as 'ESS Instrument' in Figure 7.4) in Copenhagen, Denmark. The Group Leaders I talked with worked at both sites, dividing their time between Copenhagen and Lund, and with some living in Sweden and others in Denmark. While the ESS Lund site was built from the ground up in order to house the neutron source and other parts of the experimental facility (Rekers, 2016), the ESS DMSC site was a smaller project with fewer requirements for specific space, and therefore suitable for sharing office space with other organisations.

One of the questions I had when arriving at the DMSC was what difference it would make for there to be a separate site for data management? Could

Figure 7.4: Diagram from the original technical design report for ESS showing which parts of the data management work will take place at each site

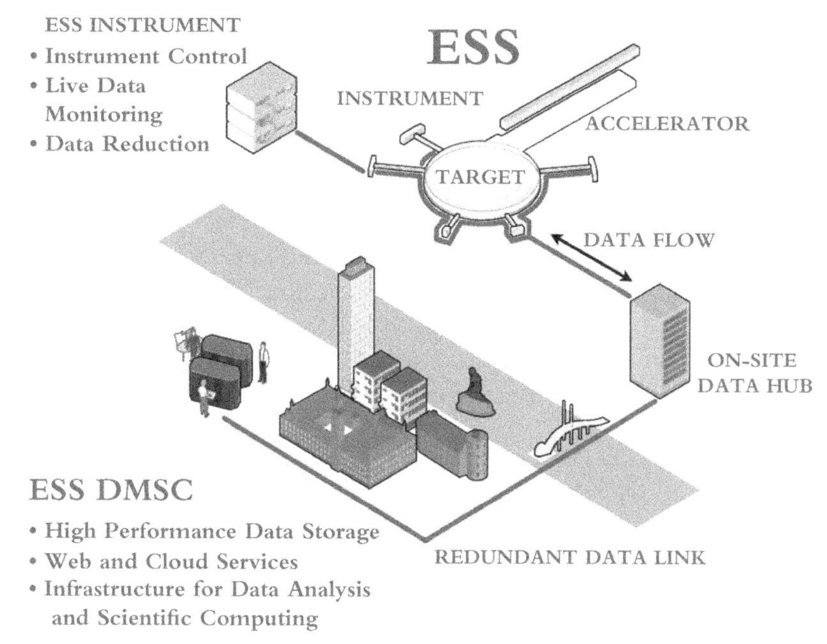

Source: Åberg et al (2013)

this herald a new era of visibility for data management? The political weight attributed to locating the DMSC in Copenhagen in recognition of the Danish contribution to the ESS has – as I have suggested elsewhere in this book – the sense that the DMSC was a significant enough part of the organisation to constitute such a public recognition. Read alternatively, it could simply be that the DMSC was the only part of the organisation considered sufficiently detached from the experiment to be able to be detached physically from the main facility. Understanding how the relationship between the two sites played out in reality was thus a key part of understanding the value and visibility accorded to data management. Petra Aulin (initially DMSC Project Coordinator and later Group Leader) was very much focused on the organisational aspect of the DMSC and in our first conversation she described some of the more practical aspects of setting up this office, which was intended as she puts it, to be an 'external little campus':

> to be this external little campus that DMSC is. One challenge is that over in Lund the planners can grab the line managers whenever they walk across the corridors, right and say, OK we need to discuss the budgets, to discuss when will you hire and so on, to get the budget to fit. But that's not really the case here because the line managers

> are here most of the time. So my role most of the time is actually just being a possibility for the line managers to come to me and we sit down and plan the budget. I go to see the planners and feed them with the information and the other way, exactly, exactly. So that's taking up a bit of my time, and then I'm still implementing a lot of the administrative routines from ESS in Lund. … So they have, for example to have a logistics routine. So how do we do when we get packages in and how do we, practical things, logistics, or you know. How do we greet our guests without getting people to go into parts where they're not allowed … you know, basic stuff. Important there but maybe not as important here. So, we had to, you know discuss, is this something we should be part of or do we need to have our routine. Is it important for ESS in Lund that we are part of their routine. You know, it's a little bit to get the feeling that we're still one company, right? (Interview with Petra, 2015)

Petra's role involves acting as a kind of bridge between the two sites, conveying expectations and needs from one to the other, as well as taking care of practical, security arrangements. However, even in this first conversation, there is a sense of different priorities emerging across the two sites ('Important there but maybe not as important here'). Petra was realistic and diplomatic in her account, acknowledging that the difference in size of site played an important role: 'it doesn't come naturally for them to think about the 15 people here, when they've 400 people there' (Petra, 2015). Her insights into the everyday processes of the DMSC illustrate the challenge of ensuring that the DMSC based in Copenhagen can function independently in some way, while remaining well-connected to the ESS Lund office in terms of organisational culture ('to get the feeling that we're still one company').

Despite such efforts to bridge the gap, the earlier conversations with the Group Leaders reflected a certain amount of frustration with a lack of control over their own work environment. This was potentially aggravated by organisational changes. This is Petra talking about how her status as part of the 'Science' group meant that she was excluded from 'Admin' meetings, despite that being the focus of her work.

> But then there was a new head of Admin. … It's not a year ago. Maybe it's a little bit less, I don't know when she started. And she has a new way of doing it where she didn't find it as a good solution to have me in that managers meeting, because number one I was not part of Admin. I'm paid out of the Science budget, and number two I'm not up at that level. So when that shifted, that collaboration for my sake, well I've lost that a bit actually. So it's a little bit harder to learn stuff. (Interview with Petra, 2016)

By the time I visit Petra for the second time in 2016, her access to ESS Lund planning had been reduced. Here the reasoning behind the shift is particularly interesting when considering the status of the DMSC. The first reason given for why Petra is excluded from administrative meetings is because she is 'paid out of the Science budget'. Within the organisation then, the DMSC (and all its staff) appears to be categorised as 'Science'. The second reason concerns Petra's management level. At the time of the interview, she was the highest ranked person at the DMSC taking care of practical organisational matters, although her job title did not match that of the administrative managers at the ESS Lund site. We can see two kinds of categorisation in play here – a distinction between Science and Admin, and a distinction between different levels of staff. Both of which are deployed strategically to exclude Petra from meetings in Lund and require extra work on her part to bridge the gap.

Meanwhile, this 'gap' between sites also affected some of the technical work, as Sune discussed:

> we just needed one file to be changed on one of their servers. … And we put this up and said, you need to change this … and nothing happened, and you know, we've been pinging them six or seven times. And we said, we can do it if you give us access to the computer. We can do it. Nah, no, we will do it. … And six months later and they haven't been able to change that file. (Interview with Sune, 2017)

Sune's account suggests a more complicated hierarchy between the two sites. While the ESS Lund site holds the permissions to change the files that Sune needs updating, there seems to be a question as to whether the delay relates to power games or technical ability. In effect, access to servers functions as a virtual boundary wall. The relationship between the two sites is expressed in multiple contexts, and whether by ignoring the needs of the DMSC, or seeking to control it through compliance with administrative routines or technical hierarchies, there is a clear sense in the earlier conversations that the DMSC is afforded a lower status or made less visible in the organisation[1]. However, it is equally important to note the change in this relation that happens over time and through staff changes.

[1] When I returned to discuss the previously discussed findings with the DMSC Group Leaders in 2019, they were keen to point out that many of the difficulties they had identified in the interviews had now been resolved due to improved communication between the sites, as well as the ESS project as a whole becoming more mature. Many of the struggles they experienced to set up the DMSC may well have been mirrored in other departments at the ESS as everyone tried to find their feet in the fast-growing organisation.

Perhaps most significant in the interviews were the changes at management level. At the time of the 2016 interviews, the Head of DMSC at that time, Mark Hagen, had just resigned. These interviews are thus very rich in reflections around what that will mean for the DMSC. Mark had many years of experience at Big Science facilities but also a management style that was very different to the person who took over after he left, Jon Taylor. In addition to this change within the DMSC, there were some major staff changes within the ESS itself, such as new CEOs. The later interviews are characterised by a growing awareness of the need to be proactive in building good relations between the sites. By the last year of interviews, Jon had formally been installed as Head of DMSC and his personal leadership style had begun to influence the relationship between the two sites: 'I think the other thing that has changed a lot is that the relationship to ESS headquarter or institute, I mean has improved significantly. Jon is simply the type who spends more time in Lund. … And that has actually meant a lot' (Thomas, 2017).

Jon's change in role meant a redistribution of his Group, as well as changes to the way the DMSC internal communication took place and the communication between the two sites. This also coincided with Petra's role being formalised as a Group Leader: 'so now I'm a group leader on the same level as the other group leaders. I am not a doctor doctor professor, but I am a line manager for people' (Petra, 2017). Together with increases in staff numbers, it is clear that these years were extremely turbulent ones for the DMSC. Some of these political tensions were also played out in concrete ways as the DMSC moved into new, permanent premises.

New office

The office space for the DMSC is based in Denmark in order to reflect the Danish contribution to the ESS project, to provide jobs, and to facilitate knowledge exchange with researchers at the University of Copenhagen. For this reason, from its inception, the DMSC has had offices in Copenhagen, rather than Lund. There was, however, a long period of uncertainty about where exactly in Copenhagen the DMSC would ultimately be housed, resulting in it having several temporary homes (mostly in University of Copenhagen premises). During the time that I was meeting my participants the DMSC offices moved to a new location within Copenhagen which is hoped will be its permanent home. This move was discussed at length during the interviews and I also took the opportunity to take photos of the two sites when visiting (included in the Introduction to this chapter). This search for a 'proper' home can be understood as an important part of making the DMSC visible both literally and symbolically (Suchman, 1995).

Achieving a permanent, stable location for its work was an important part of the way in which the DMSC was to gain legitimacy in the wider organisation and scientific community. By virtue of being in a separate location to the source, the DMSC is made more visible organisationally and politically. It also creates a number of practical challenges to be addressed. In the first round of interviews Petra highlighted that an important part of her work at that time was finding new premises for the DMSC. Subsequently, almost all of the interviews conducted in 2016 made reference to the new office building which the DMSC had by then occupied for a few months.

> So it's not smooth here, and the fact that we don't have a kitchen to sit in does not really make you feel settled in. It's still like you've moved into some half construction site thing where you've got your bed in the corner, but when you want to have food, it's the chippy round the corner. … So it's, feeling really settled in is a bit difficult. (Interview with Tobias, 2016)

> I tried to make it as smooth as possible, and I have tried to involve people and ask them, you know how they wanted to be, or how could we do it better, and you know, they wanted curtains and we arranged for curtains and so on. But it's not the dream, it's not the dream scenario and of course I wish that I could make something. (Interview with Petra, 2016)

The staff increases that occurred drove not only the move from one office to another, but many of the disgruntled feelings that seem to be connected to the new premises. Petra is responsible for organising the work environment, and a recurrent theme in our conversations was the move to the new premises. After the move there was a lot of dissatisfaction among DMSC staff about the space. While Petra clearly wanted to improve this for her colleagues, her hands were tied because the budget for office space for all ESS departments, including DMSC, was held in Lund: 'So when it comes to the office space, we have quite a challenge, because it turns out it's too small. So it's crowded. People hate it. So and we need to expand, and the budget for expanding is in admin in Lund' (Petra, 2017). Jon added, 'It turns out not everybody likes working in an open plan office. Some people really don't like it' (2017). In Petra's comment she focuses on the issue of office space to illustrate the challenge for a geographically separate and now quite sizeable unit to lack control over some key aspects of its daily running. Meanwhile, Jon's account focuses on the organisation of the space and how that connected to the working environment.

What had looked initially like a move to a very much more visible and professional workspace separate from the university was proving problematic, and it became clear that DMSC had limited power to change this. While some people were unhappy with the open plan arrangement of the space, more pressing practical issues were apparent too. The DMSC had expanded so fast that they had already run out of space at the new offices. Furthermore, the server room for handling all of the experimental data – a key part of the DMSC responsibilities had still not been installed at the new office and the budget for doing this was under negotiation. Planning the space and requirements for this key resource also gave some clues as to how data management was understood by other members of the organisation.

Formal planning for the server room required conversations with other groups within the facility as to the quantity of data involved. This revealed a widespread lack of awareness about the role of data management but also provided an opportunity for raising awareness of the consequences of user expectations. For example, Sune described some of the conversations he was having about how much data will be processed and how this will impact the size of the server room:

> you get all these different kind of numbers, some say ten to the sixth, some say ten to the seventh, ten to the eighth, ten to the ninth and there's a big difference between a million and a billion you know, a second that is. So there's a big difference there and for me there's a question of, should I have, you know a 20 square metre server room or a 2000 square metre server room. (Interview with Sune, 2015)

Sune's experience reflects both a lack of awareness around the infrastructural requirements for management of large data sets, and also hints at a broader organisational lack of knowledge. By being located in a separate premises/city/country, highly specific practical questions about data management, such as the size of the server room, became visible and articulated in ways that might not have happened if DMSC was part of the same site and data management was rolled in with other scientific computing needs.

In the long term, the DMSC is likely to remain too small a unit to have its own building, although this has been discussed as a possibility. Therefore, it is probable that it will continue to share office space. The discussion around the move from university to commercial premises was interesting to observe in that it lifted up several concerns, namely the importance of balancing maintaining a visible connection to the university, with developing a distinct, separate identity for the DMSC. Petra's comments also suggest a balance being negotiated between DMSC as a distinct organisational unit and as being part of a larger ESS collective identity with shared practices. While relations between the two ESS sites may not always have been easy, during the three years of interviews it is possible to see the DMSC developing a clearer identity of its own. The office move occurred in tandem with the move from concept to design to prototype with several of the data management technologies. It also took place in parallel with a recruitment drive that saw nearly all of the groups expand or even double in size.

The new premises – by virtue of the trouble it caused – may also have had the effect of making the DMSC more visible across the wider ESS organisation. From struggles over budget to driving conversations about the size of data sets in order to plan the server room, the questions forced an attention to the DMSC that contributed to its slowly emerging identity organised around a particular aspect of the experimental process.

The In-Kind Contribution model

While internal negotiations about premises constitute an important entry point for discussing the organisational context in which DMSC has been developing data management solutions, equally important has been the extensive external network with whom the DMSC is in conversation. The principal formal mechanism for this has been the IKC model, through which other European countries can provide equipment, software and expertise (see also Chapters 1 and 5). The ESS website boasts that the scale on which the IKCs are taking place is unprecedented for a Big Science project: 'Thirteen Founding Member nations, around 40 European in-kind partners and more than 130 global institutions are all cooperating to build the world's next-generation neutron source' (https://europeans pallationsource.se/in-kind-contributions (accessed 24 July 2023)). The financial importance of these contributions across the ESS cannot be underestimated: 'These IKCs are expected to finance more than 35% of the total €1.843 billion (2013) construction costs' (https://europeanspall ationsource.se/ess-organisation).

For each one of these contributions, there is a specific contract and delivery schedule, constituting a significant piece of organisational workload and administration. For the DMSC the impact of these agreements is even more significant as they constitute a larger percentage of their budget: 'So we are aiming for 60, 65% in kind, right, of our budget' (Petra, 2015). How then do the IKCs fit into the organisational context of the DMSC?

In my initial conversations with Jon and Tobias, the IKCs seemed to build on their existing network of contacts as well as an existing spirit of scientific cooperation. Here, for example, is Tobias talking in 2015 about how he approaches the IKCs for his group:

> We want to have something that works for us but we don't want it to be something where they just say, ok we'll just write this code. They told us to do this, right so we'll just slave away and then once the project is over we will just throw it over the fence and be done with it, right. We want this to be something that they have a proper stake in and

> self-interest that we can also use it here … we have to be all friends with each other and motivate them in the way that scientists are usually motivated by making something work. (Interview with Tobias, 2015)

Here the IKCs seem to be particularly well-suited to the spirit of scientific discovery, as distinct from a commercial environment where the emphasis is on competition rather than collaboration. This spirit of friendly cooperation persisted throughout my conversation with Tobias, as for example in the following year when he connected this model to a broader benefit for the scientific community: '(t)hree of the large sources now work together in ways they partially probably could have done for decades' (Tobias, 2016). Furthermore, he identified an additional benefit to the ESS that is slightly harder to quantify but also valuable in terms of organisational development:

> there is also a benefit that we get out of this in kind thing that, I mean you don't have to line manage these people, like they are embedded in an organisation that knows how to run a facility, and they just do sensible things, because they're embedded in this organisation that knows how to do stuff. And we still need to learn to do that at times. (Interview with Tobias, 2016)

Learning from other organisations' experience of setting up structures and work processes is framed here as an additional benefit that Tobias recognises as being particularly helpful in the early stages of ESS development. His reflection also suggests how organisational practices may be shared or spread between organisations. The spirit of scientific cooperation no doubt facilitates such sharing of experience and contributes to the emergence of the ESS organisational culture. In Chapter 5 I noted how experiences from other facilities shaped the choice of equipment (in the case of Kafka), and in Chapter 6 I highlighted how the Group Leaders' backgrounds at other facilities played an important role. Here it is possible to see how friendly working relations also shape organisational aspects of the ESS.

For some of the Group Leaders the impact of the IKCs is enormous. However, for others much less so:

> I think on the software side there is sort of a, there's a, there's something to be benefited from having experts delivered in kind because you can sort of reach out and get inspiration from people who are already working in the field. For the hardware side, it's more like, it's a project given, you know. You need to have this amount of money done in kind, so we will work together with somebody to procure the hardware. But really, there's not much gained from that. (Interview with Sune, 2015)

Sune explains that part of the IKC for his group comprises provision of servers by a partner, but his experience of this process is much less like a collaboration. Meanwhile for Thomas, the expertise his group needs lies more within the scientific community itself ('a lot of the software that we provide has been developed by a scientist' (Thomas, 2015)) meaning that he adopted a more flexible model that included IKCs as part of other collaborative working practices, such as code camps.

In Thomas' account, he describes trying to balance the need to access specific expertise which often lies with the scientists themselves who develop their own software as a side project, with an organisational need to provide long-term stability: 'So my task is really in many cases, to secure the software in five or ten years from now for these people right, and try to go in and negotiate with some of these people behind the software and things like that' (Thomas, 2015). Thomas is aware that when individuals develop their own software, this creates a potential risk for a facility that adopts that software as there is then a single point of failure if the developer retires or leaves the field. His job involves bridging the gap between present and future users, as well as moving from individual development to a potentially more stable, standardised form of the software: 'I try to set up collaboration agreements and we try to set up the infrastructure, so we have sort of like some level of quality assurance' (Thomas, 2015). His narrative thus connects strongly with the question of how data management within Big Science is changing, epitomising as it does themes of standardisation, professionalisation and increased user support.

It is clear that the different requirements for each group (hardware vs software), and the existing working practices within the broader community (individual development by scientists vs standardised/commercialised solutions) shape how they adopted and adapted the IKC model to suit their goals. However, when I listen closely, I also wonder to what extent the Group Leaders own backgrounds shaped how they used the IKC model. Both Jon and Tobias came to ESS with backgrounds and extensive contacts within Big Science, and no experience of commercial software development. Their accounts suggest that they used the IKC model to develop existing networks framed within a specific sense of how *scientific* collaborations work. Meanwhile, Thomas and Sune brought their experience and expectations of *professional* development with them, and used the IKCs differently.

In this section, I have suggested how the IKC model has been interpreted through the lens of the individual Group Leaders' experience and in response to the specific demands of the pieces of the technical puzzle results. As such it provides a useful example of how personal experience, organisational structure and technical challenges are intertwined at the DMSC. The IKCs were developed as a way to address the enormous price tag associated with the ESS and function both as a practical solution to constructing the facility and as a way of performing particular power relations. In the following section, I turn my attention to the budget itself as an important site for organisational negotiations.

Budget

Earlier in this chapter I suggested that one of the things that makes DMSC more visible is its physical location separate from the ESS Lund site. By virtue of being in Copenhagen the DMSC cannot help but be more visible, appearing as a separate organisational unit with its own budget. This setup could mean that the DMSC would – unlike historical arrangements for scientific computing – be better supported in having its own defined budget. This did not mean, however, that the budget for the DMSC was without controversy – at least during the time that I was following the Group Leaders. Budget discussions often represented a moment in our conversations when infrastructuring practices became more visible, and particularly the relations between DMSC and other parts of the ESS organisation. For this reason, this section explores what can be learnt about organisational practices and politics from a close reading of my participants' reflections around the budget. Two topics that occurred repeatedly were the rebudgeting exercise in 2016 and the administrative budget for the DMSC.

During the fieldwork period the ESS underwent a rebudgeting exercise in which the entire facility reviewed its budgets, and certain activities were moved from one construction phase into another. This exercise showed which activities needed to be prioritised in order for the ESS to keep as closely as possible to its planned opening schedule, and which items could be left until a little later for completion. This exercise not only had practical implications for the DMSC in terms of planned staffing, for example, but also became a point of discussion in terms of how it reflected the status of the DMSC within the organisation. Outgoing Head of DMSC, Mark Hagen, put this neatly when he said:

> When people originally, you know, a few years back they could see all this happening and said, OK, the Danish thing of having a DMSC was, OK you've got to do that in advance. But you know, there is a reversion to type recently, you know. Oh we don't need the software until we need it, you know. We've got to spend the money on other things, you know. (Interview with Mark, 2016)

In Mark's comment that 'there is a reversion to type' we can hear an acknowledgement of the way in which data management in Big Science has previously been neglected, and perhaps some disappointment that – despite promises to the contrary – data management at the ESS seemed to be starting to suffer from the same problem. Mark expresses a sentiment that I heard across several of the conversations, which was that funds for the DMSC were not felt to be prioritised compared to other parts of the ESS. While some of the Group Leaders were pragmatic about the need to put funds into, for example, instrument development, others seem to feel that this was shortsighted. Thomas painted a particularly vivid picture of the problems that could be caused by neglecting to invest in data management:

> I've heard about beamlines where they produce maybe two publications per year, right. It's nothing, and the cost of it is the most expensive publications in the world, right [LAUGHS]. And it's basically because the users can't figure out how to interpret the data. Something like that. Less than, it's hard to find the number for this, but you know some of the users' facilities, apparently a lot less than half of the data are actually published. So there's a huge waste. (Interview with Thomas, 2016)

Thomas' reflection highlights the consequences of what happens if data management is not properly supported and makes clear why it is important not to neglect this part of the experimental pipeline. Like Mark he draws attention to the bigger picture of how data have been approached and understood in other Big Science facilities historically. In both Thomas and Mark's accounts, there seems to be hope that the DMSC might be different. If the 'big promise' (and justification for the hefty price tag) of Big Science is to produce knowledge that changes our understanding of the world, but data from the experiments is not published then it becomes easier to question the usefulness of Big Science facilities.

While the rebudgeting exercise dominated primarily the 2016 conversations, a theme that was present throughout the three years of conversations concerned the administrative budget for the DMSC. While the majority of activities carried out by the DMSC fell under the so-called 'Science budget' (for which DMSC had its own budget), a number of more practical aspects such as working space had been allocated to the 'Administrative' budget. The administrative budget covering premises was held centrally in Lund, resulting in the DMSC having limited control over their working environment. As Petra's comments in the earlier section on the new office indicate, personal relationships between DMSC staff and ESS Lund staff were key here. Her account of the difficulty in planning for the DMSC acknowledges that part of this may be attributed to the difference in size between the two ESS sites: 'So they're also having an extremely tight budget, and I would guess. This is only my guess, but we are 22 people in Denmark and they are 450 people in Sweden. They have bigger problems. Or they see it as bigger problems, right?' (Petra, 2017).

Petra recognises that the different size of the DMSC site compared to the ESS Lund site plays a role in how the budget is allocated but also in affecting which parts of the organisation appear to take precedence. This illustrates one of the ways in which DMSC was limited in its capacity to shape its own work environment and also suggests a power imbalance between the two sites that once again frames DMSC as a supporting event to the main show. Those parts of the budget held in Lund at the central ESS administration involved delicate negotiations to ensure that the resources required by the DMSC were not overlooked. Petra's diplomacy and Jon's work to build good relations between the sites can be viewed as essential emotional labour to smooth organisational politics and protect the interests of the DMSC. At times, this intersection of personal and organisational impacts directly the data management technologies, as was in the case of the server room.

During the rebudgeting exercise, it became clear that some key aspects of the DMSC's activities were simply not included in the budget. One example of this concerns the server room that is an important part of the DMSC premises: 'So there was no money in the budget to actually do any installation of any server in the basement. This was missed. It should have been in something. … Half a million Euros should have been budgeted for. Anyway, it wasn't as it turned out' (Jon, 2017).

Jon's blunt narration of the lack of funds for the DMSC server room once again had me wondering about the gap between rhetoric and practice that Mark had discussed in the previous year's interviews. The public discourse around the DMSC had played up its importance in terms of both acknowledging the centrality of data management in the experimental process, and the importance of the Danish contribution to the ESS project. However, as the project progressed, emerging omissions in the organisational planning such as the funds for an essential piece of DMSC equipment undermine this data-positive storyline. Zooming in on the budget allows us to glimpse one way in which organisational friction plays out and power relations between the two sites are enacted.

Conclusions

In this chapter I have drawn attention to organisation as an important part of the context surrounding big data at the ESS but also gestured to how the case of the ESS may indicate changes in how Big Science approaches data management. I follow Eric Nost in suggesting that 'data is infrastructured not through technical means per se, but through the ways experts manage fiscal and institutional "frictions" to data integration and maintenance (Baker and Karasti, 2018; Bates, 2017; Edwards, 2010)' (Nost, 2022, 106). This chapter, then, has focused on three specific areas of 'friction' as moments when data management becomes more visible. I opened this chapter by describing the two very different kinds of office premises occupied by the DMSC during the period of my fieldwork. The new office was one of those areas of friction, closely connected to the question of budgets (another kind of friction). The remaining 'friction' that I explored was that of the IKC model. Here the friction is not about internal tensions at the ESS, but rather how varying deployment of the IKC model reflects tensions between 'scientific' compared to 'commercial' models for development. The DMSC is made visible thanks to the increasing need for expert data management in Big Science and the location of the unit itself as physically separate from the experimental facility. However, the underlying power structure within the organisation, as evidenced most often through budgeting issues, suggests that underlying assumptions about the role and status of data management in Big Science are slow to change. The specific examples of premises, collaborations and budget give a flavour of the negotiations and practices that are ongoing. The reflections of my participants also show a high degree of ambivalence over the course of the three years, with the importance of data management becoming more or less visible subject to organisational pressures. While moments of organisational friction give the DMSC opportunities to be visible inside the organisation, these are temporary and, as the infrastructure at least appears to settle, I wonder if the DMSC will become less visible over time?

If you want good data, then there is a cost for that. Something that these frictions make abundantly clear. The cost may be calculated either in terms of someone's time (for example, the hours spent by the postdoc producing home-made analysis software in the past) or in money (commercial software 'bought in'). The IKC model deployed by the ESS represents a transitional mode, in which a combination of models can be seen, with contributions taking the form of expertise, programming or equipment. Such a model makes it possible for enormous collaborative efforts like the ESS to take place, and highlights the importance of building such facilities as part of an existing network of technical and scientific communities that span many countries. Inevitably such collaborative endeavours demand negotiation, but most importantly personal skill. In focusing on organisational friction, this chapter also lifts up the invisible work done by the Group Leaders to carry out such negotiations, both inside the DMSC, across the ESS organisation, and throughout the international IKC model.

8

The Rawness of the Data

When I was preparing this book, one of the reviewers encouraged me not to fall into the trap of assuming that the data at the European Spallation Source (ESS) was 'cooked' already but rather to keep an open mind, to consider the possibility that the data I was studying might be truly raw. Intrigued and somewhat provoked by this, this chapter is an attempt to step back a little and explore how the theoretical provocation offered by Critical Data Studies (CDS) might be used as a question that could be posed to the conversations I had with the Group Leaders (rather than as a 'fait accompli'). This would mean framing CDS scholarship as offering one possible critical perspective among many – one possible truth rather than The Truth about data in the ESS. Viewed through the lens of CDS the previous three chapters highlight the journey undergone by the data during experiments – through affordances/limitations of hardware and software, through the varied professional experiences of the Group Leaders, and through organisational priorities and practices. However, what also became clear in my interviews, is that the Group Leaders don't necessarily share my view that this journey affects the 'rawness' of the data.

The first part of this book's title (*Behind the Science*) acts here as both metaphor and provocation. Big Science is a show with an international audience, lots of competition and a big budget. By invoking the feel of a stage show with lots going on behind the scenes as well as on stage, the metaphor of 'behind the science' is useful because it acknowledges this visibility, this performance, and invites us to wonder what might be going on behind the big discoveries that change our worlds. The metaphor of in front and behind the stage curtains permits two understandings of the data to coexist. Front stage we have the 'magic' of the experiment itself led by teams of international scientists seamlessly producing (ta-da!) results to widen our eyes, meanwhile back stage (behind the scenes, or should that be behind the science?) the mechanics of the show occur where the hardworking production crew members tinker, plan, hold things together and know the internal workings. This metaphor becomes provocation when we start to explore how the promised transparency and objectivity of scientific experimentation grounded in 'raw' data might also be something of an illusion. The previous chapters suggest ways in which data might be 'cooked' through the mechanics of data management behind the scenes. However, this claim also risks upsetting the correlation between 'raw' data and valid knowledge production, also known as objectivity.

Data management in Big Science is changing, as the case of the Data Management and Software Centre (DMSC) illustrates. What then does it mean when the 'behind the scenes' activities change – how might that impact Big Science? Under what circumstances might the production crew move out of the shadows of behind the scenes and become visible as part of the science? What are the stakes involved in including data management formally within The Experiment (for Science, for scientists and for the data management experts themselves that I met)? The critical lens offered by the black box and invisible work help here to shine the spotlight on the work of the Group Leaders and to look inside some of the technical black boxes they have been constructing and aligning to ensure the data flow.

I started from the premise that part of what made 'New Big Science' distinctive was the increasing size and complexity of the data sets emerging from experiments being carried out at these facilities, and the changing face of visitors to such facilities. The two combined (data and users) have resulted in the emergence of increasingly formalised support around data management for users, thus heralding the emergence of a new professional group. This formalisation and visibility prompts questions about how such work is recognised and valued but also about the very nature of the data that is being produced. These technical and professional shifts have here been placed into a particular scholarly context in which critical perspectives on big data have been gathering pace and ferocity for more than a decade already. A scholarly context characterised by the maxim 'raw data is an oxymoron and a bad idea'. However, it would be too easy and oversimplistic to present this study as a CDS exercise in (ta-da! and here we draw back the curtain that normally hides the back stage work …) 'revealing' the ways in which data will be 'cooked' by the data management procedures, practices and technologies at the ESS. I am more interested in exploring what my own and my participants' fidelity to our own paradigms and practices does to our understanding of Big Science.

In this chapter, then I will examine how differing understandings of 'raw' data emerge in my participants' reflections, and consider how these may point to wider changes in Big Science. In so doing, I will attempt to take a step back and keep an open mind. What impact on the data journey do the processes, contexts and perspectives that I detailed previously have? Can the data still be considered to be 'raw'? What is at stake in drawing a line about when and where the data are 'raw' in the context of Big Science in general and the ESS in particular? Answering these questions involves homing in on the 'rawness' of the data being produced as a means to explore objectivity, transparency and Science.

Why the data needs to be raw

In this section, I return to the Group Leaders to explore how they understood and explained 'raw data' to me and how this shifted over the three years of interviews. I include extracts from the three years, which I read with careful attention to how they change and point us towards the importance of context when defining what counts as 'raw' data.

> From the moment the first neutrons produced by the European Spallation Source (ESS) register their existence on a detector, the raw experiment data will flow from Lund, Sweden, to Copenhagen, Denmark, and then on to the ESS scientific user community throughout Europe. (https://neutronsources.org/news/news-from-the-neutron-centers/pan-european-ess-data-management-and-software-centre-takes-form-in-copenhagen/)

This quotation appeared as the opening lines in a press release that appeared in September 2015, just weeks before I was due to start interviewing the Group Leaders for the first time. This press release is illustrative in that it describes the data both as 'raw' and as having their existence registered, a rhetorical move which presents the detector as the moment of what Leonelli might call 'data birth'. A lengthy piece, it also provided useful background reading for me as I prepared my questions. I asked each of the Group Leaders to explain how their area of responsibility connected to the information provided in the press release as a way to understand their different responsibilities:

> which I feel responsible for in that sense, is the part which says, streams the raw data. So it's the first part of the sentence. Because stream means I need to set up a network that can handle the data rates that the instruments are providing. So I need to be able to take the data as they happen and make sure they're available as they happen. (Interview with Sune, 2015)

> in principle, I mean it is just when raw data comes in and then there is some pipeline with, I don't know, three to thirty or something like that different data treatments. (Interview with Tobias, 2015)

> It's getting from what we call raw data, something about you get some bits coming out of some detectors, binary, I don't actually know what it is ... and then they convert that to physical data. Physical data is,

> you have for instance a wavelength of your incoming neutrons on the one axis and intensities of your data. (Interview with Thomas, 2015)

I include these extracts from three of the Group Leaders to illustrate how 'raw data' was used in the first year of interviews. With the exception of Thomas, they were all relatively recent arrivals at the DMSC and were engaged in the very early stages of planning how data management would work. Throughout 'raw data' is used to describe data that occurs early in the pipeline, without any detail or caveats on what that meant. There is an acknowledgement from all of them that there is processing that takes place ('treatments', 'convert that to physical data'). In 2016 I returned to do a second round of interviews and found more nuancing in the way that 'raw data' was being used. In the following extract I am talking with Jon, who is showing me a data set he had generated as a way to explain to me:

Jon:	It's my data.
Katherine:	So you've just generated a bunch of data?
Jon:	I took it on an instrument. I should maybe show you, actually this is. So this is, I've just loaded a raw data file and its numbers.
Katherine:	And these are the numbers we were talking about?
Jon:	This is an Excel spreadsheet. This is not event mode data, but the principle is the same. And then ... the idea is this is what the instrument looks like in real life. It's a, these are individual detector tubes on an instrument called MARI which is installed at ISIS where I used to work. And there's 900 of them. (Interview with Katherine and Jon, 2016)

When Jon says 'in real life' he imbues these numbers with authority by claiming that when we look at these numbers we see exactly what the detectors see during an experiment. There is a directness and an immediacy to the turn of phrase that is powerful. For Jon, then, raw data is 'event mode data', or the string of numbers that tell the scientists the relevant time and space information, such as the speed at which the neutron travelled or in which direction it moved. These numbers mean very little on their own, before they have been transformed into what Thomas referred to as 'physical data' in his earlier quotation, meaning something that resembles wavelengths. If we go back and look once more at Jon's quotation, we might say that 'event mode data' sounds pretty 'raw' ... right? There has been no transformation of this into 'physical data' and no analysis of the results. However, if I look closely at my conversation with Tobias during the same year, there are clues that Jon's data may already have undergone a process in which it is 'refined':

Katherine: So you've done a first cut of it?
Tobias: We have basically just refined it to the bare minimum of the raw data. It doesn't, we don't know the wavelength of the neutron that can, that needs to be calculated later depending on other parameters that are then also in Kafka. (Interview with Katherine and Tobias, 2016)

Even before the transition from event mode data to physical data, the data undergo a process of 'refinement' which means:

> It's processed detector data which means it, just the raw data, the raw information that you want from the detector. The detector delivers more information, or when capturing a single neutron, you get a number of signals. So we get a number of signals but we just want that to be refined to-, we had a neutron here. (Interview with Tobias, 2016)

Tobias uses the term 'raw data' in a very specific way, which he is careful to explain to me. In this quotation he is describing the data that has come from the detector and which has been 'refined' meaning that several results for the same neutron are reduced to one result, a process which makes data more manageable. It has not been shaped in any other way. The data here is 'raw' from the perspective of the visiting scientists looking for a result in their experiments. To me, however, this process of refining the data contains within it a number of assumptions about what constitutes 'valid' data and therefore how several data points might be refined into one result. Those assumptions are not clear here and – for me with my critical hat on – trouble the 'rawness' of the data, although not for Tobias. Tobias and Jon seem to have the same notion of when 'raw data' occurs (somewhere after the string of binaries produced by neutrons hitting detectors, but before Kafka (see Chapter 5) transforms event mode data into physical data).

By year three of the interviews, further details emerge that suggest disciplinary differences are also important to our discussions on raw data. Jon makes this clear when we are discussing the need for supercomputing in order to work with experimental data:

> So there are some experiments which you, to do anything with the data, even to understand the data, you need some kind of model and most of the time some simulation. And there are other experiments where you can do analytical simulations … and parameterised model of the model system. I mean the idea is I suppose that the, the data itself isn't really, I mean OK if you go away from things like maybe diffraction where you can absolutely tell where the atomic positions

> are and the raw data … and the crystal symmetry. If you go to neutron scattering or material science, you need a model. You need to do a good publication. You need to understand the data and be able to communicate what that understanding is. So you need some kind of, yeh simulation to understand what is actually happening inside the material. (Interview with Jon, 2017)

Jon's explanation highlights how different techniques work more or less closely with the 'raw data'; when using diffraction there is no requirement for processing of the data in order to be able to 'read' it, but neutron scattering requires a model 'to understand what is actually happening'. When talking later that same year with Tobias I asked to what extent users are aware of the different kinds of processing that take place. He explained that one of the goals for the DMSC is to develop a data catalogue that makes more visible the different versions of the data set – for example, raw, reduced, calibration:

> and you can also automate processing for other techniques. Which isn't always done. You want to do that for most things. But then even if it is done, it's not necessarily in the metadata catalogue as in, this is the raw data and here, look, you don't really need to look at the raw data. Here is the processing result. And that really changes how people perceive doing their science. (Interview with Tobias, 2017)

For Jon, Tobias and the others data is 'raw' at a particular place and time in the system. For users this may be a different place, and they may not be aware of the processing that has taken place which encompasses calibration of the detectors, refinement of the event mode data, or processing for particular techniques. With the goal of building a data catalogue that makes visible the different data sets, the DMSC could be seen to be working towards a more transparent view of data management that offers the user a window into the process and the impact on their data. Implicit in this is an understanding that many things may affect the data.

Over the course of the three years my return visits allowed me to gain a more nuanced understanding of how the term 'raw data' is understood and used by the Group Leaders. The nuances showed the importance of context. 'Raw' data (as I had imagined it) is a series of zeroes and ones, although for Jon and Tobias at least it seems like event mode data is closer to 'raw'. However, it is not very useful to visitors in that form, requiring as it does transformation into physical data, and then potentially also processing with a model in order to make sense of it. In brief, it entirely depends on what you want to use the data for, and how your scholarly/scientific community understands 'valid' data.

Gitelman (2013) points out that the minute we start thinking about data we have already framed it within our own disciplinary context, epistemological paradigms and knowledge of available and appropriate methods, thereby 'situating' it before even conducting an experiment. For the Group Leaders, these framings are different, shaped of course by their own professional and intellectual backgrounds, but also by the pragmatic realities of organisational life, and the material limitations of the technologies themselves. Thus the understanding of what is 'raw data' could be quite different for the Group Leaders compared to visiting scientists:

> So if you start by laying out everything that you would possibly need to do and some of the things are done behind the scenes. You calibrate the detectors and then the user just uses the data that comes off the detector. They don't really know that this is happening. (Interview with Tobias, 2017)

Much of the work (although not all) that the Group Leaders do takes place 'behind the scenes' and that inevitably changes their view on what constitutes 'raw' data. Users 'don't really know this is happening' and for me that underlines the metaphor of 'raw data' as an illusion, an illusion that obscures the work done by people and machines behind the scenes of Big Science:

> a central part of the tasks undertaken by technicians is directed towards making machines run smoothly, so that the creation of usable data is unproblematic. They must make the working of laboratory apparatus and procedures faultless and invisible so that scientific work can be published and can thus count as knowledge. The more successful they are, the less visible they – and their work – becomes. (Iliffe, 2008, 5)

Iliffe argues that the machines and the people must disappear in order for the data to be 'raw'. At least, that is how Science has traditionally handled data. However, perhaps things are changing? Tobias' description of what the metadata catalogue should contain hints perhaps at increasing transparency around how data might be processed during the very early stages of its life cycle and in ways not always directly connected with human contact.

What counts as 'raw' data seems – in essence – to depend very much on where you are standing (epistemology), and what you consider to be the best way to answer your question (methodology). The earlier extracts also suggest differences between those whose role is more technically oriented and those who are used to working with data to answer scientific questions, not to mention the audience. The nuances in the conversations show how the term shifts to adapt to specific uses and contexts, but also the lure of the term 'raw data' as a way to create validity.

Situated rawness and objectivity

Hopefully in the previous section I have persuaded you that what counts as 'rawness' of data is highly context specific, and that the ways in which the Group Leaders use this term reflects that. What then is at stake when we use the term 'raw data' in the context of Big Science? What would it mean to make the journey undertaken by the data, and its effects upon the data, more clear? Tobias' interest in developing a metadata catalogue that brings together the various different kinds of data related to an experiment feels like a step in this direction – at least to me – and points to a greater transparency in the data management process. What if these changes with data are illustrative of changes in science more broadly? Could this be indicative of a sea change in how 'raw' data is correlated with objectivity? Are we entering a new era of objectivities? After all, it would not be the first time that the measure of objectivity has shifted. Peter Galison, for example, notes in *Image and Logic: A Material Culture of Microphysics* that:

> Roughly in the middle of the nineteenth century, a change occurred, and a visual notion of objectivity emerged that broke significantly with the older conception of truth to nature. Now the goal was not to reconstruct, idealize, or approximate the form behind a wealth of actual objects, but rather to exert super human effort to remove the author from the process of depiction. (Galison, 1997, 121)

Perspectives from CDS suggest that even our fundamental assumptions about what 'data' might be for us constitute a framing or situating of our experiments. If we adopt this stance, then our expectations of what data we are trying to collect are a fundamental part of this 'Science'. But equally that our understandings of our Science shape the kind of data we wish/expect to collect. Science and Data are inextricably intertwined in the quest for objectivity. As Gitelman notes in her summary of Daston and Galison's work on 'objectivity': 'The point is not how to judge whether objectivity is possible – thumbs up or thumbs down – but how to describe objectivity in the first place. Objectivity is situated and historically specific' (Gitelman, 2013, 4).

As Big Science facilities become slowly more interdisciplinary, it could be useful to reflect on how the foundational understandings of what is 'data' in Big Science derive from specific disciplinary traditions and practices (historically High Energy Physics or HEP). By opening up facilities to other kinds of users, what other kinds of epistemic paradigms might come into play in understanding what 'data' are necessary for that discipline's notion of 'objectivity' and how to calibrate the machine accordingly? This would be in line with the broader attention to how data management technologies are changing how Science is done. These technologies – by dint of their open or closed software, their collaborative building processes, their ability to share data across multiple institutions – play a role in shaping the careers of individual researchers, contributing to institutional collaborations or tensions, or even changing how we understand what it means to do research:

> current databasing and networking efforts cross disciplinary and institutional boundaries, and the integration of tools in these new research practices may mean that the distinction between types of tools (say, between databases and models) and between activities (like representing and analysing) may be difficult to draw. (Beaulieu, 2001, 636–7)

Beaulieu, for example, sees digital technologies as fundamentally changing the production of knowledge through the convergence of different functions and forms into one medium. Beaulieu's account, which focuses on the rise of neuroinformatics, details the rationale behind the development of the field in the 1980s. Driven by the agenda of the Institute of Medicine in the US, the development of large databases of neuroscience information were not only envisaged as an opportunity for modelling but also as a way to avoid duplication of information and facilitate translation of research into clinical medicine. These latter two reasons reflect the role of data management in collaboration between institutions and disciplines.

Writing in *Structures of Scientific Collaboration*, Shrum et al examined several collaborations using as a starting point data in the AIP Center for History of Physics (2007). This was supplemented by interviews with scientists, policy makers and funding agencies to track their histories of organisational collaboration around physics, particularly on the collaborations that form around accelerators. The authors examined the conditions that determined scientific collaborations, including people, agreements, data sharing and publications. They asked: Is it a big topic with lots of scope to produce lots of data for many people to work on? Will there be enough discrete projects to satisfy the career ambitions of all, especially junior scholars needing to find their niche? And how is authority and validity of the data guaranteed – through formalisation of the collaboration in the form of documentation showing who collects what and how, or through hierarchy in which all the collected data is sent to a single point for checking and analysis? In their work we can see how the structure of the research team is entangled with the practices of data collection, and the creation/maintenance of 'valid' data.

Beaulieu and Shrum et al offer just two examples of how Science is changing through/thanks to changing data practices. Are these changes also moving us towards a changing notion of objectivity or are they new ways of achieving objectivity as it has been understood in the most recent centuries? The concepts 'situated knowledges' and 'strong objectivity' coined by feminist Science and Technology Studies (STS) scholars and which have gained wider traction in the last 30 years are characterised by efforts to do precisely the opposite of what Galison describes. Including the context around knowledge production is now considered best practice in some disciplines, indicating transparency, accountability and recognition of the diversity of human experience.

Seams and alignment

In the preceding section, I leaned on Galison, Gitelman, Beaulieu and Shrum et al to remind us of a bigger picture, one in which objectivity and Science are (i) intimately entangled with 'raw' data and the technologies used to collect it and (ii) sociohistorically specific in how they are achieved. Understanding the 'rawness' of data as being a notion which is highly flexible, available for leveraging in different ways and intimately entangled with bigger notions of 'objectivity' returns us once again to the idea of 'situated knowledges'.

In pointing out the ways in which the data management process is 'situated' in specific contexts (organisation, technology and people), my reading aims to contribute to CDS scholarship (see Chapter 2) by problematising the 'rawness' of the data being produced. The diagram I sketched with my participants (Figure 5.2) shows the different technologies through which the data passes on its journey from detector to scientist. Each station on the data journey provides an opportunity to dig into the particularities of what happens to data at that point (for example, what happens to the data when it goes into that box? And what can you tell me about where that box came from? What might go wrong here?). And every time that investigation is performed a particularity emerges. It reveals the context around selection of that technological artefact and how it might have been otherwise, as well as exploring how the decision that was made had a ripple effect on the surrounding relationalities and on the data itself.

My exploration of the different technologies used at the DMSC through the lens of CDS deliberately plays with different understandings of what constitutes 'raw' data, in a context where the expectation from users of the facility is of receiving 'raw' data direct and in real time from the experiment. In the preceding sections I have shown how 'raw data' is by no means a stable notion and is itself subject to different understandings and uses, whether we look closely at the DMSC itself and the differing definitions of raw data proffered by my participants or at the bigger historical picture in which objectivity is shaped by changing technological paradigms. In the preceding chapters I have suggested how specific instances/moments/technologies in the data management process might situate the data.

What then is at stake in suggesting that the data that will be produced at the DMSC is 'always, already cooked'? To help me answer this last question, I want to return us to the discussion I began in Chapter 5 about alignment work. There I explored Jon's comment that the big challenge for the DMSC team was in 'making all the bits work together' through Janet Vertesi's use of alignment (2014). Focusing on alignment work made it possible to lift up the invisible work done by the DMSC to sew together the 'seams' between various people and machines that move data from detector to scientist, and through this to 'situate' the data produced as the result of a particular data journey. While focusing on the seams allowed me to situate the data, for the users of the facility it will be of utmost importance that the seams do not appear and the data remains 'raw'.

A seamless data journey allows the data to remain 'raw', it creates an illusion of transparency in the system. Good alignment is thus also essential to preserving the boundary line between 'raw' and 'cooked' data upon which the validity of the experiment rests. Like 'rawness', transparency is a quality that is produced through a set of very particular performances and procedures. Its aim is to guarantee a certain quality or reliability to the data (if we return to the metaphor of the stage, this would be the moment where someone says 'look, nothing up my sleeves!'). While the 'rawness' of the data is offered up as an indicator of the truthfulness of the data itself, transparency typically says something about the organisation or process that supplies the data. Thus, transparency is about trust in the process, and rawness is about trust in the product (the data). Both aspects are increasingly in demand both in scientific research and society more widely, perhaps most notably in relation to calls for open data (Denis and Goëta, 2014). Transparency, like 'rawness' has also increasingly been the target of scholarly critique which challenges its promise of objectivity (Michener and Bersch, 2013; Amoore, 2020).

Seamlessness is important in the context of the ESS to ensure real-time flow of enormous quantities of data and to maintain trust in the experimental procedure by providing reliable 'raw' data to scientists (Kruse, 2006). In order to smooth the seams not only must different components and people be aligned, also the data itself must be in a format that ensure a smooth journey through the system. For my participants, then, 'seams' (meaning the neatly stitched together boundaries of different teams or different pieces of equipment) are both resources to be exploited (opportunities to form collaborations, find new solutions) and the evidence of problems solved. A seam means that two boundaries/two edges have been joined. The better the alignment work, the more seamless the join, the more invisible my participants' work becomes and the more transparent the system, thereby 'guaranteeing' raw data – at least for the users, if not the developers, or the CDS scholars.

> Technicians' work was transparent when the apparatus was working as it should and the results were as they ought to be. In contrast, the role of technicians was continually pointed to when matters did not proceed as expected. (Shapin, 1989, 558)

> The users, that's the beauty of all of this, they shouldn't see any of that almost. If the users really know about the details of that, they're either very curious and bored and talk to people that really don't have much input on the experiment, or it's some catastrophic failure that shouldn't happen, and we need to explain why it happened. (Interview with Tobias, 2016)

At a local level, seamlessness is an important part of the way in which 'raw' data is produced and managed, and a part of maintaining a good professional reputation for the DMSC. At a more metaphorical level, seamlessness is part of creating the illusion, maintaining the magic of objectivity. However, a seamless performance not only disappears the various forms that the data takes, it also disappears the DMSC behind the scenes of the Main Show of the ESS. In the particular context of data production at a Big Science facility, zooming in on the work done to align seams between technologies and people produces a sensitivity to shifting conceptualisations of rawness during the journey taken by scientific data produced during experiments, and perhaps an increased awareness of the stakes of visibility. To whom must the seams be invisible in order to guarantee a transparent process and 'raw' data? For whom are well-stitched seams a source of professional pride (Kruse, 2023) and evidence of work well done? When does the 'situatedness' of the data flicker in and out of sight, and why?

Conclusions

I have discussed previously the notion of invisible work in relation to data management in Big Science. The invisibility of this work was what inspired the title of this book (*Behind the Science*), because the work of data management feels like it takes place behind the scenes of the main show of Big Science.

Working within a scientific paradigm in which raw data is the prerequisite basis for experimental results that can provide an objective answer to a question, it matters a great deal that the data continues to be described as 'raw' so that the experimental results may stand alone as a reliable, reproducible understanding of a world. For some paradigms, then, there is a great deal at stake in poking a finger at the 'rawness' of data. And noting the specificity of that paradigm is key. However, as my version of the story suggests, this is not the only paradigm. Other scientific communities may think differently, and – as the users of Big Science facilities evolve, as well as standards for data management – maybe some of these paradigms will encroach upon Big Science. This is expensive data to produce: maybe it's right that society should have a clearer picture of how it is produced. Maybe it's useful to know that this kind of data management is becoming more standardised and professionalised, and to start to ask if this makes a difference? Regardless of whether we consider the data 'raw' or not, I'd like to think that these are questions upon which my participants and I could agree.

9

Big Science: A Moving Target

Which organisations do you think of when someone says 'big data'? Facebook? Amazon? In fact, the largest quantities of data produced in today's data-dominated world occur as the result of experiments taking place at Big Science facilities. At these facilities, experiments using powerful sources produce vast amounts of data and lead to results that fundamentally change our understanding of the world around us. It's puzzling then that we know so little about how such data is collected and managed at these facilities.

This book was written out of a project that set out to explore the idea that Big Science itself was changing in significant ways, so much so that we might need to start talking about 'New' Big Science. My contribution to that conversation was to point out that perhaps one of the distinguishing characteristics of such a New Big Science was a change in data – the volumes of data, the complexity of the data, the changes in user support around data, all of which combined to make data management at facilities such as the European Spallation Source (ESS) more visible and potentially more valued than ever before. Of course, this is set against the much larger backdrop of global discussions that started with big data, and moved swiftly on to open data, data security, data privacy, passing through General Data Protection Regulation (GDPR), and emerging with artificial intelligence (AI) and machine learning (ML) related concerns about data sets used to train AI and the potential of synthetic data. This high visibility of data in both research and popular culture is not new. It is, however, an important part of the bigger conversations around how data might be understood in Big Science because it indicates a pressing need across the board for more critical conversations about the relationship between data and knowledge production.

In the Introduction to this book, I outlined the puzzle that hooked me from the start: Why – when data is so foundational to experimental results – does the management of that data receive so little attention in Big Science? This was confirmed in my early conversations with the Group Leaders of the Data Management and Software Centre (DMSC), as Thomas explained:

> it's interesting to see how these extremely expensive facilities, how little effort is put into software development and how to manage it. It's amazing. You were saying you spend a billion or 1.8 in the case of ESS and you think that you would be in control of your software development and computers and things like that. But no, that's not the case. And even though you will think, well in order to get output out of this facility, you need to have all these things running and in good shape and all that. The last thing people should meet is IT issues or the computers are not working, or you can't analyse your data, things like that. But that's how it is actually. (Interview with Thomas, 2015)

In order to understand why this was the case, I took a tripartite approach to examining data management at the ESS, in which I focused on people, technologies and organisation. Given the absence of accounts of data management in the Science and Technology Studies (STS) scholarship on Big Science, the concepts of invisible work and the black box early on presented themselves as useful thinking tools. While both are attuned to shifting in/visibilities, they also provide complementary ways of examining both daily practices and the development of technologies. The central chapters of this book thus focus on technologies, people and organisation in turn. Each chapter uses the critical attention to in/visibility provided by the concepts of invisible work and the black box to analyse the conversations I had with the Group Leaders of the DMSC. The DMSC is the most visible manifestation of the need for expert support around data management in Big Science to date. This is both by virtue of the challenges and debates listed previously but also – somewhat pragmatically – because of the political jockeying around the location of the ESS as part of European-wide negotiations on hosting this ground-breaking facility that led to it being located separately from the neutron source. These changes in visibility emerged in a variety of ways across my conversations.

In Chapter 5, the focus is on the implementation of a key piece of the data management process, called Kafka. The journey taken by the data from detector to scientist constitutes a kind of 'black box' where the input is the experimental sample and the output is a set of numbers on a screen. This journey involves various pieces of hardware and software, most of which is invisible to users. Thus a major challenge for my participants is 'making all the bits work together' so that the journey is seamless. Here Kafka plays an important role in ensuring safe movement of the data. In this chapter I took inspiration from Janet Vertesi's work about alignment to pose questions to the process my participants described, bringing their relationships and processes into focus as activities where humans work together to stitch the seams between technologies. Alignment work is a way to lift up the invisible work done by the DMSC to align various people and machines to move data from detector to scientist. Listening to the descriptions of how Kafka was selected as the technical solution for this work allowed me to open up the 'black box' of data management by looking at data in movement – the metaphors of journey and friction work here as a way to pay attention to this movement and the inevitable difficulties/risks involved in such movement.

Chapter 6 shifted the focus from the technology to the people, zooming in on my participants' own stories about their career trajectories. Here I paid attention to their own use of the notion of what constitutes 'professional' approaches to development of data management systems. This allowed me to explore both their own backgrounds before working at the ESS, as well as their perceptions of the structures and practices in place at the ESS. Differences in experience highlighted an important question about whether my participants are technical or scientific staff, and the implications of this boundary line for experimental practice. Unlike scientists, technicians can remain invisible and out of the experiment, thereby allowing them to work with the data and the data still to remain 'raw' when it reaches the screens of the visiting scientists. The downside to this is that data management and scientific computing has received little institutional support in Big Science historically. The DMSC potentially represents a change in this status with increased visibility raising the possibility of a change in professional validation. In the final part of this chapter, I asked what is at stake in the professionalisation of my participants' work, putting this into dialogue with the debate about 'raw' data.

The topic of professional software development appeared also in the third and final strand which focuses on organisation (Chapter 7). The move from a university building to commercial premises, coupled with the In-Kind Contribution (IKC) model of development, prompted many reflections during my conversations about how the ESS compared to other organisations in terms of support for data management. Zooming in on the negotiations around working space, I explored how visible and invisible infrastructuring practices (from budgets to server rooms to cross-site dynamics) indicated ambivalences about the status of the DMSC within the wider organisation. Moments where there was organisational tension (or friction) meant that negotiations were more visible and available for reflection. Meanwhile, the IKC model represented a topic through which relations to external networks and communities could be explored, revealing a high level of specificity in how the model was deployed depending on what part of data management was in focus.

Overall, these chapters show how the DMSC is geographically, organisationally and intellectually separated from the 'main show' of the experiment, which will take place in Lund. This separation is both product and producer of the invisibleness of the work done in scientific computing, and which is achieved through creation and maintenance of certain boundaries. These include what might be called 'intellectual' boundaries such as technician/scientist that determine who is allowed/positioned in a way to do 'Science' publicly. It also includes organisational boundary lines such as unit budgets and membership of administrative groups, allowing us to see more clearly how power relations are enacted through everyday practices. Studying these boundary lines brought me to a point where I started thinking about experiments at the ESS as having a 'front stage' and a 'back stage' aspect to them, a metaphor that inspired the title of this book and which I develop further in Chapter 8.

Chapter 8 is the place where I engage most explicitly with the scholarly discussions taking place in Critical Data Studies (CDS) which seek to deconstruct the notion of 'raw' data through contextualising this as the product of epistemological, methodological, technical and sociocultural conditions, as Lisa Gitelman explains: 'Every discipline and disciplinary institution has its own norms and standards for the imagination of data, just as every field has its accepted methodologies and its evolved structures of practice' (Gitelman, 2013, 3).

If we accept Gitelman's argument, then we might productively use 'raw data' as a way to ask: for whom, under what conditions and with what motivations does something need to be called 'raw data'? From this starting point, we can use the concept of 'raw data' as a way to examine the limits and assumptions of scientific paradigms, as well as to enquire into the changing status of data management in Big Science. In this chapter I develop the idea of 'Behind the Science' to connect the rawness of data with the invisibility of both the technologies themselves and the work being done by people at the DMSC.

Chapters 5 through 7 are thus intended to provide a glimpse into an aspect of Big Science that is currently under-researched in the STS and CDS literature. Chapter 8 is intended to contribute to an ongoing theoretical discussion around critical perspectives on data and data management. Together they seek to respond to the following questions:

- How do epistemological assumptions about data, material limitations of the technical equipment and organisational politics intersect to shape the data that will be collected at a new Big Science facility?
- What role does data management play in a broader landscape of changing scientific experimentation as Big Science becomes New Big Science?

The ambitious vision for the DMSC's role shown in the diagram at the start of Chapter 5 reflects the increased demand for technical support from non-expert users, as well as a response to a much broader ongoing conversation across facilities and users about the future of scientific computing. With this reflection in mind, there are two things to highlight here. First, the visibility of the work being done by the DMSC changes. Much of the work of the DMSC takes place 'behind the scenes' of what the users see (as the title of this book suggests) but it is front stage for others. Their expertise in designing server rooms, connecting pieces of disparate software to ensure that the information about the experimental environment is attached to the detector results is an essential part of making ESS work but not one that is visible to visiting scientists unless it breaks. Inevitably, perhaps, the 'customer' for the services of the DMSC was therefore not always the visiting scientist conducting an experiment. In my conversations with participants, a variety of people appeared as stakeholders or conversation partners in their discussions about which software or hardware to deploy. Tobias talked about IKC collaborators, while Thomas talked about user communities represented through Scientific and Technical Advisory Panels (STAP). Noting these differences helped me to see that while some of the decisions they made were 'behind the scenes' for visiting scientists, these were often the decisions that formed an important part of ongoing conversations and knowledge-exchange with peers at other facilities.

Second, at each stage of this diagram there are multiple pieces of hardware and software engaged in delivering on this promise. While some of these are independent, much of the system must function together so that a proposal for an experiment is connected to the sample when delivered, and then to the experiment. In other words, behind this neat diagram is a complex technical system that is engaged with running an experiment, some parts of which are more focused on data than others but which are nonetheless interconnected. Thus, the data is one part of a much bigger complex system, comprising both material assets and virtual ones. The material assets such as the ESS premises may command more immediate attention – if only by virtue of their literal size and the disruption caused. The virtual assets such as data produced or expertise harnessed may be harder to 'see'. My sketched diagram in Chapter 5 (Figure 5.2) works to try and make some of these harder-to-see assets more visible.

There are layers upon layers of development histories and contexts contained within Figure 5.2 that show the data journey from detector to scientist. Studying the DMSC turned out not to be as simple as 'being there before the box closed' because the box was full of other boxes developed in other places, and by other people, some of which were connected to the Group Leaders (for example, Jon's long history with Mantid) and some which were not (for example, Kafka). This resulted in varying levels of opacity to the different components in the data management journey. Each component reflects certain assumptions and limitations that surrounded its development, while the particular alignment of the components reflects the skills and networks of the Group Leaders and their teams.

Paying attention to the specificity created by these layers helps us to understand what is special about Big Science. Part of the opening gambit of this book was that there is something rather special about 'Big Science' as a context for studying data management. Understanding data management in this context as a blend of technical and scientific expertise, commercial and amateur software development, external collaboration and internal negotiation situates the development of the data management systems at the ESS as the result of a specific time, place and group of people. I opened this book by enquiring into why data seems to be so invisible when it is the foundation of the knowledge production emanating from Big Science, and, related to that, if changes to data management are symptomatic of changes in Big Science. What I hope has become clear through this book is that 'Big Science' is a very much a moving target and always has been. As a number of the commentators mentioned in Chapter 2 noted, attempts to define this scholarly endeavour are notoriously tricky. But what then of the data? As experimental practice has evolved so too has the data, in format, volume, complexity. Thus the techniques and technologies for handling such data have also inevitably evolved, resulting in a multi-stage journey to be undertaken by the data. Following this journey made visible the different kinds of work taking place to support each stage. This work included development of individual 'black boxes' but also the work to align them, work that is most visible during construction or repair. With this book, I hope to contribute an account of how data handling at the ESS will take place, an account which seeks to build on those important accounts of earlier data management practices by making visible the context in which this work takes place and the importance of individual expertise.

References

Åberg, M., Ahlfors, N., Ainsworth, R., Alba-Simionesco, C., Alimov, S., Aliouane, N. and Alling, B., 2013. *ESS technical design report*. European Spallation Source. https://europeanspallationsource.se/sites/default/files/downloads/2017/09/TDR_online_ver_all.pdf

Akrich, M., 1992. The de-scription of technical objects. In W.E. Bijker and J. Law (eds) *Shaping technology/building society: Studies in sociotechnical change* (pp 205–24). MIT Press.

Amoore, L., 2020. *Cloud ethics: Algorithms and the attributes of ourselves and others*. Duke University Press.

Antoniou, A., 2021. What is a data model? An anatomy of data analysis in high energy physics. *European Journal for Philosophy of Science*, 11(4), p 101.

Apache Kafka. https://kafka.apache.org/ (accessed 29 June 2024).

Baker, K.S. and Karasti, H., 2018, August. Data care and its politics: Designing for local collective data management as a neglected thing. In Proceedings of the 15th Participatory Design Conference: Full Papers, Volume 1 (pp 1–12).

Bates, J., 2018. The politics of data friction. *Journal of Documentation*, 74(2), pp 412–29.

Bates, J., Lin, Y.W. and Goodale, P., 2016. Data journeys: Capturing the socio-material constitution of data objects and flows. *Big Data & Society*, 3(2), p 2053951716654502.

Beaulieu, A., 2001. Voxels in the brain: Neuroscience, informatics and changing notions of objectivity. *Social Studies of Science*, 31(5), pp 635–80.

Bietz, M.J. and Lee, C.P., 2012. Adapting cyberinfrastructure to new science: Tensions and strategies. In *Proceedings of the 2012 iConference* (pp 183–90).

Blaauw, A., 1988. ESO's early history, 1953–1975. I. Striving towards the convention. *The Messenger*, 54, pp 1–9.

Blok, A., Nakazora, M. and Winthereik, B.R., 2016. Infrastructuring environments. *Science as Culture*, 25(1), pp 1–22.

Bokulich, A., 2020. Towards a taxonomy of the model-ladenness of data. *Philosophy of Science*, 87(5), pp 793–806.

Bokulich, A., 2021. Using models to correct data: Paleodiversity and the fossil record. *Synthese*, 198(Suppl 24), pp 5919–40.

Borgman, C.L., 2015. *Big data, little data, no data: Scholarship in the networked world*. MIT Press.

Bowker, G.C., 2008. *Memory practices in the sciences*. MIT Press.

Bowker, G.C. and Star, S.L., 2000. *Sorting things out: Classification and its consequences*. MIT Press.

Buckland, M.K., 1989. Information handling, organizational structure, and power. *Journal of the American Society for Information Science*, 40(5), pp 329–33.

Calvert, J., 2013. Systems biology, big science and grand challenges. *BioSocieties*, 8, pp 466–79.

Cetina, K.K., 1999. *Epistemic cultures: How the sciences make knowledge*. Harvard University Press.

Couldry, N. and Mejias, U.A., 2019. Data colonialism: Rethinking big data's relation to the contemporary subject. *Television & New Media*, 20(4), pp 336–49.

Couldry, N. and Mejias, U.A., 2020. *The costs of connection: How data are colonizing human life and appropriating it for capitalism*. Stanford University Press.

Cramer, K.C., 2017. Lightening Europe: Establishing the European synchrotron radiation facility (ESRF). *History and Technology*, 33(4), pp 396–427.

Cramer, K.C., 2020. *A Political History of Big Science*. Springer International Publishing.

Cramer, K.C. and Hallonsten, O. eds, 2020. *Big science and research infrastructures in Europe*. Edward Elgar Publishing.

Crease, R.P. and Westfall, C., 2016. The new big science. *Physics Today*, 69(5), pp 30–6.

Crease, R.P., 1999. *Making physics: A biography of Brookhaven National Laboratory, 1946–1972*. University of Chicago Press.

Daniels, A.K., 1987. Invisible work. *Social Problems*, 34(5), pp 403–15.

Dalton, C. and Thatcher, J., 2014. What does a critical data studies look like, and why do we care? Seven points for a critical approach to 'big data'. *Society and Space*, 29. https://www.societyandspace.org/articles/what-does-a-critical-data-studies-look-like-and-why-do-we-care

Denis, J. and Goëta, S., 2014. Exploration, extraction and 'rawification'. The shaping of transparency in the back rooms of open data. *The Shaping of Transparency in the Back Rooms of Open Data*. After The Reveal. Open Questions on Closed Systems – Neil Postman Graduate Conference, February 2014, New York, United States (28 February 2014).

Dimitrievski, I., 2019. *Accounting the future: An ethnography of the European Spallation Source* (Vol. 771). Linköping University Electronic Press.

DMSC. 11 April 2023. https://europeanspallationsource.se/data-management-software-centre (accessed 29 June 2024).

Doing, P., 2008. Give me a laboratory and I will raise a discipline: The past, present, and future politics of laboratory studies in STS. *The Handbook of Science and Technology Studies*, 3, pp 279–95.

Doing, P., 2009. *Velvet revolution at the synchrotron: Biology, physics, and change in science*. MIT Press.

Edwards, P.N., 2010. *A vast machine: Computer models, climate data, and the politics of global warming*. MIT Press.

Edwards, P.N., Mayernik, M.S., Batcheller, A.L., Bowker, G.C. and Borgman, C.L., 2011. Science friction: Data, metadata, and collaboration. *Social Studies of Science*, 41(5), pp 667–90.

Ehrlich, K. and Cash, D., 1999. The invisible world of intermediaries: A cautionary tale. *Computer Supported Cooperative Work (CSCW)*, 8(1–2), pp 147–67.

ESS, 2005. https://neutronsources.org/news/news-from-the-neutron-centers/pan-european-ess-data-management-and-software-centre-takes-form-in-copenhagen/ (accessed 1 July 2024).

ESS, 2009a. 'Agreement over european research facility, ESS'. University of Copenhagen, Niels Bohr Institute website: http://www.nbi.ku.dk/english/news/news09/research_facility_ess/ (accessed 28 May 2015).

ESS, 2009b. 'In-Kind Contributions'. https://europeanspallationsource.se/in-kind-contributions (accessed 27 June 2024).

ESS, 2015. https://neutronsources.org/news/news-from-the-neutron-centers/pan-european-ess-data-management-and-software-centre-takes-form-in-copenhagen/ (accessed 1 July 2024).

ESS Activity Report, 2022. https://pub.mediapaper.se/15fcdfe7-10ba-430e-ae98-558f177bf337#page=3 (accessed 12 October 2023).

ESS, 2023. https://europeanspallationsource.se/data-management-software-centre (accessed 14 February 2023).

ESS Computer Centre, 2024. https://europeanspallationsource.se/datamanagement-software/computing-centre (accessed 27 June 2024).

Gahl, T., Hagen, M., Hall-Wilton, R., Kolya, S., Koennecke, M., and Rescic, M., 2015. *Hardware aspects, modularity and integration of an event mode data acquisition and instrument control for the European Spallation Source (ESS)*. arXiv preprint arXiv:1507.01838.

Galison, P., 1997. *Image and logic: A material culture of microphysics*. University of Chicago Press.

Garfinkel, H. and Bittner, E., 1967. Good organizational reasons for 'bad' clinic records. *Studies in Ethnomethodology*. Prentice Hall.

Garforth, L., 2012. In/visibilities of research: Seeing and knowing in STS. *Science, Technology, & Human Values*, 37(2), pp 264–85.

Garg, N., 2013. *Apache Kafka*. Packt Publishing.

Gehl, R.W., 2016. The politics of punctualization and depunctualization in the digital advertising alliance. *The Communication Review*, 19(1), pp 35–54.

Gillespie, T., Boczkowski, P.J. and Foot, K.A. eds, 2014. *Media technologies: Essays on communication, materiality, and society*. MIT Press.

Gitelman, L. ed, 2013. *Raw data is an oxymoron*. MIT Press.

Glymou, B., 2000. Data and phenomena: A distinction reconsidered. *Erkenntnis*, 52(1), pp 29–37.

Haider, J. and Kjellberg, S., 2016. Data in the making: Temporal aspects in the construction of research data. In *New big science in focus: Perspectives on ESS and MAX IV* (pp 143–63). Lund Studies in Arts and Cultural Sciences, Lunds universitet.

Hallonsten, O., 2012. *In pursuit of a promise: Perspectives on the political process to establish the European Spallation Source (ESS) in Lund, Sweden.* Arkiv förlag & tidskrift.

Hallonsten, O., 2013. Introducing 'facilitymetrics': A first review and analysis of commonly used measures of scientific leadership among synchrotron radiation facilities worldwide. *Scientometrics*, 96(2), pp 497–513.

Hallonsten, O., 2014. How expensive is Big Science? Consequences of using simple publication counts in performance assessment of large scientific facilities. *Scientometrics*, 100(2), pp 483–96.

Hallonsten, O., 2016. *Big science transformed.* Palgrave Macmillan.

Hanseth, O. and Monteiro, E., 1997. Inscribing behaviour in information infrastructure standards. *Accounting, Management and Information Technologies*, 7(4), pp 183–211.

Haraway, D.J., 1991. Situated knowledges: The science question in feminism and the privilege of partial perspective. In *Simians, Cyborgs and Women: The Reinvention of Nature* (pp 183–201). Free Association Books.

Haraway, D.J., 1997. *Modest_Witness@Second_Millenium. FemaleMan©_Meets_OncoMouse™. Feminism and Technoscience*. Routledge.

Harris, T., 2003. Data models and the acquisition and manipulation of data. *Philosophy of Science*, 70(5), pp 1508–17.

Harrison, K., 2015. (Gendered?) histories: Tracing the development of an ICT in the Swedish rescue services, in A. Brammé (ed) *Yearbook 2013 of the Institute for Advanced Studies for Science, Technology and Society*, Profil Verlag GmbH.

Heilbron, J.L. and Seidel, R.W., 1989. *Lawrence and his laboratory: A history of the Lawrence Berkeley Laboratory* (Vol. 5). University of California Press.

Heidler, R. and Hallonsten, O., 2015. Qualifying the performance evaluation of Big Science beyond productivity, impact and costs. *Scientometrics*, 104, pp 295–312.

Hermann, A., Krige, J., Mersits, U. and Pestre, D., 1987. *History of CERN* (Vol. 1). Amsterdam: North-Holland.

Hine, C., 2006. Databases as scientific instruments and their role in the ordering of scientific work. *Social Studies of Science*, 36(2), pp 269–98.

Hoddeson, L., Kolb, A.W. and Westfall, C., 2019. *Fermilab: Physics, the frontier, and megascience*. University of Chicago Press.

Hood, L. and Rowen, L., 2013. The human genome project: Big Science transforms biology and medicine. *Genome Medicine*, 5, pp 1–8.

Iliffe, R., 2008. Technicians. *Notes and Records of the Royal Society*, 62(1), pp 3–16.

Kaiserfeld, T. and O'Dell, T. eds, 2013. *Legitimizing ESS: Big Science as a collaboration across boundaries*. Nordic Academic Press.

Karaca, K., 2020. What data get to travel in High Energy Physics? The construction of data at the large hadron collider. In S. Leonelli and N. Tempini (eds) *Data Journeys in the Sciences*, pp 45–58. Springer Nature.

Karasti, H. and Baker, K.S., 2004, January. *Infrastructuring for the long-term: Ecological information management*. In 37th Annual Hawaii International Conference on System Sciences, 2004. Proceedings of the (10 pp). IEEE.

Karasti, H. and Blomberg, J., 2018. Studying infrastructuring ethnographically. *Computer Supported Cooperative Work (CSCW)*, 27, pp 233–65.

Karasti, H. and Syrjänen, A.L., 2004. Artful infrastructuring in two cases of community PD. In Proceedings of the eighth conference on Participatory design: Artful integration: interweaving media, materials and practices (Vol. 1, pp 20–30).

Karasti, H., Baker, K.S. and Millerand, F., 2010. Infrastructure time: Long-term matters in collaborative development. *Computer Supported Cooperative Work (CSCW)*, 19, pp 377–415.

Kirtz, J.L., 2018. Beyond the blackbox: Repurposing ROM hacking for feminist hacking/making practices. *Ada: A Journal of Gender, New Media, and Technology*, (13). https://doi.org/10.5399/uo/ada.2018.13.3

Kitchin, R., 2014. *The data revolution: Big data, open data, data infrastructures and their consequences*, Sage

Kitchin, R. and Lauriault, T., 2014. *Towards critical data studies: Charting and unpacking data assemblages and their work* (July 30, 2014). The Programmable City Working Paper 2; pre-print version of chapter to be published in Eckert, J., Shears, A. and Thatcher, J. (eds) Geoweb and Big Data. University of Nebraska Press. Forthcoming, Available at SSRN: https://ssrn.com/abstract=2474112

Krige, J. ed., 1996. *History of CERN* (Vol. 3). Elsevier.

Kruse, C., 2006. *The making of valid data: People and machines in genetic research practice (Doctoral dissertation)*. Linköping University Electronic Press.

Kruse, C., 2023. Swabbing dogs and chauffeuring pizza boxes: crime scene alignment work and crime scene technicians' professional identity. *Science & Technology Studies*, 36(4), pp 62–79.

Latour, B., 1983. Give me a laboratory and I will raise the world. In Karin D. Knorr-Cetina and Michael Mulkay (eds) *Science observed: Perspectives on the social study of science*, pp 141–70, Beverly Hills, CA: Sage.

Latour, B., 1987. *Science in action: How to follow scientists and engineers through society*. Harvard University Press.

Latour, B., 1999. *Pandora's hope: Essays on the reality of science studies*. Harvard University Press.

Leonelli, S., 2015. What counts as scientific data? A relational framework. *Philosophy of Science*, 82(5), pp 810–21.

Leonelli, S., 2019. *Data-centric biology: A philosophical study*. University of Chicago Press.

Leonelli, S. and Tempini, N., 2020. *Data journeys in the sciences*. Springer Nature.

Lievrouw, L.A., 2014. Materiality and media in communication and technology studies: An unfinished project. In T. Gillespie, P.J. Boczkowski and K.A. Foot (eds) *Media technologies: Essays on communication, materiality, and society* (pp 21–51). MIT Press.

Lydahl, D., 2017. Visible persons, invisible work? Exploring articulation work in the implementation of person-centred care on a hospital ward. *Sociologisk forskning*, 54(3), pp 163–79. https://doi.org/10.37062/sf.54.18213

Lykke, N., 2016. Intersectional analysis: Black box or useful critical feminist thinking technology?. In H. Lutz, M.T.H. Vivar and L. Supik (eds) *Framing intersectionality: Debates on a multi-faceted concept in gender studies* (pp 207–20). Routledge.

McAllister, J.W., 1997. Phenomena and patterns in data sets. *Erkenntnis*, 47(2), pp 217–28.

Marcheselli, V., 2020. The shadow biosphere hypothesis: Non-knowledge in emerging disciplines. *Science, Technology, & Human Values*, 45(4), pp 636–58.

Michener, G. and Bersch, K., 2013. Identifying transparency. *Information Polity*, 18(3), pp 233–42.

Murillo, L.F.R., Gu, D., Guillen, R., Holbrook, J. and Traweek, S., 2012. Partial perspectives in astronomy: Gender, ethnicity, nationality and meshworks in building images of the universe and social worlds. *Interdisciplinary Science Reviews*, 37(1), pp 36–50.

Nielsen, K.A. and Svensson, L., 2006. *Action research and interactive research: beyond practice and theory*. Shaker Publishing.

Nost, E., 2022. Infrastructuring 'data-driven' environmental governance in Louisiana's coastal restoration plan. *Environment and Planning E: Nature and Space*, 5(1), pp 104–24.

Ohno-Machado, L., 2012. To share or not to share: that is not the question, *Science Translational Medicine*, 4(165), pp 1–4.

Open Data Foundation. https://www.odaf.org/ (accessed 27 June 2024).

Paine, D. and Lee, C.P., 2020. Coordinative entities: Forms of organizing in data intensive science. *Computer Supported Cooperative Work (CSCW)*, 29, pp 335–80.

Petrick, E.R., 2020. Building the black box: Cyberneticians and complex systems. *Science, Technology, & Human Values*, 45(4), pp 575–95.

Pickering, A. and Cushing, J.T., 1986. *Constructing quarks: A sociological history of particle physics*. University of Chicago Press.

Pipek, V. and Wulf, V., 2009. Infrastructuring: Toward an integrated perspective on the design and use of information technology. *Journal of the Association for Information Systems*, 10(5), p 1.

Plantin, J.C., 2019. Data cleaners for pristine datasets: Visibility and invisibility of data processors in social science. *Science, Technology, & Human Values*, 44(1), pp 52–73.

Plantin, J.C., Lagoze, C. and Edwards, P.N., 2018. Re-integrating scholarly infrastructure: The ambiguous role of data sharing platforms. *Big Data & Society*, 5(1), p 2053951718756683.

Price, D.J.D.S., 1963. *Little science, big science*. Columbia University Press.

Rekers, J.V., 2016. How close is close enough for interaction?: Proximities between facility, university, and industry, In: *New big science in focus: Perspectives on ESS and MAX IV.* Eds. Rekers, J.V. and Sandell, K. Lund Studies in Arts and Cultural Sciences, pp 45–68.

Rekers, J.V. and Sandell, K., 2016. *New big science in focus: Perspectives on ESS and MAX IV*. Lund Studies in Arts and Cultural Sciences.

Ribes, D., 2019. STS, meet data science, once again. *Science, Technology, & Human Values*, 44(3), pp 514–39.

Rio Poncela, A.M., Romero Gutierrez, L., Bermúdez, D.D. and Estellés, M., 2021. A labour of love? The invisible work of caring teachers during Covid-19. *Pastoral Care in Education*, 39(3), pp 192–208.

Riordan, M., Hoddeson, L. and Kolb, A.W., 2019. *Tunnel visions: The rise and fall of the superconducting super collider*. University of Chicago Press.

Saez-Rodriguez, J., Rinschen, M.M., Floege, J. and Kramann, R., 2019. Big Science and big data in nephrology. *Kidney International*, 95(6), pp 1326–37.

Scroggins, M.J. and Pasquetto, I.V., 2020. Labor out of place: On the varieties and valences of (in) visible labor in data-intensive science. *Engaging Science, Technology, and Society*, 6, pp 111–32.

Seidel, R.W., 1998. 'Crunching numbers' computers and physical research in the AEC laboratories. *History and Technology, an International Journal*, 15(1–2), pp 31–68.

Seidel, R.W., 2008. From factory to farm: Dissemination of computing in high-energy physics. *Historical Studies in the Natural Sciences*, 38(4), pp 479–507.

Shapin, S., 1989. The invisible technician. *American Scientist*, 77(6), pp 554–63.

Shapin, S. and Schaffer, S., 2011. *Leviathan and the air-pump: Hobbes, Boyle, and the experimental life*. Princeton University Press.

Shindell, M., 2020. Outlining the black box: An introduction to four papers. *Science, Technology, & Human Values*, 45(4), pp 567–74.

Shrum, W., Genuth, J. and Chompalov, I., 2007. *Structures of scientific collaboration*. MIT Press.

Simoulin, V., 2017. *Sociologie d'un grand équipement scientifique: le premier synchrotron de troisième génération*. ENS éditions.

Smyth, H., Nyhan, J. and Flinn, A., 2020. Opening the 'black box' of digital cultural heritage processes: Feminist digital humanities and critical heritage studies. In Kristen Schuster and Stuart Dunn (eds) *Routledge international handbook of research methods in digital humanities* (pp 295–308). Routledge.

Stanhill, G., 1999. Climate change science is now big science. *Eos, Transactions American Geophysical Union*, 80(35), pp 396–97.

Star, S.L., 1999. The ethnography of infrastructure. *American Behavioral Scientist*, 43(3), pp 377–91.

Star, S.L. and Bowker, G.C., 2006. How to infrastructure. In Leah A. Lievrouw and Sonia Livingstone (eds) *Handbook of new media: Social shaping and social consequences of ICTs* (pp 230–45). SAGE Publications Ltd.

Star, S.L. and Ruhleder, K., 1994, October. *Steps towards an ecology of infrastructure: Complex problems in design and access for large-scale collaborative systems*. In Proceedings of the 1994 ACM conference on Computer supported cooperative work (pp 253–64).

Star, S.L. and Strauss, A., 1999. Layers of silence, arenas of voice: The ecology of visible and invisible work. *Computer Supported Cooperative Work (CSCW)*, 8, pp 9–30.

Suchman, L., 1995. Making work visible. *Communications of the ACM*, 38(9), pp 56–64.

Suchman, L.A., 2007. *Human-machine reconfigurations: Plans and situated actions*. Cambridge University Press.

Suppes, P., 1966. Models of data. In Ernest Nagel, Patrick Suppes and Alfred Tarski (eds) *Studies in logic and the foundations of mathematics* (Vol. 44, pp 252–61). Elsevier.

Traweek, S., 2009. *Beamtimes and lifetimes: The world of high energy physicists*. Harvard University Press.

Vermeulen, N., 2013. From Darwin to the census of marine life: Marine biology as big science. *PloS One*, 8(1), p e54284.

Vermeulen, N., 2016. Big biology. *NTM Zeitschrift für Geschichte der Wissenschaften, Technik und Medizin*, 24(2), pp 195–223.

Vertesi, J., 2014. Seamful spaces: Heterogeneous infrastructures in interaction. *Science, Technology, & Human Values*, 39(2), pp 264–84.

Weinberg, A.M., 1961. Impact of large-scale science on the United States: Big Science is here to stay, but we have yet to make the hard financial and educational choices it imposes. *Science*, 134(3473), pp 161–4.

Wenger, E., 1998. Communities of practice: Learning as a social system. *Systems Thinker*, 9(5), pp 2–3.

Wikipedia, 2023. https://en.wikipedia.org/wiki/Spallation_Neutron_Source (accessed 17 October 2023).

Winner, L., 1993. Upon opening the black box and finding it empty: Social constructivism and the philosophy of technology. *Science, Technology, & Human Values*, 18(3), pp 362–78.

Wylie, C.D., 2020. Glass-boxing science: Laboratory work on display in museums. *Science, Technology, & Human Values*, 45(4), pp 618–35.

Index

References to figures appear in *italic* type. References to footnotes show both the page number and the note number (231n3).